JavaScript

JavaScript
Syntax and Practices

Dr Ravi Tomar
Associate Professor, School of Computer Science,
University of Petroleum & Energy Studies,
Dehradun, India

Ms Sarishma Dangi
Assistant Professor, Graphic Era University,
Dehradun, India

CRC Press
Taylor & Francis Group
Boca Raton London New York

CRC Press is an imprint of the
Taylor & Francis Group, an **informa** business
A CHAPMAN & HALL BOOK

First Edition published 2022
by CRC Press
6000 Broken Sound Parkway NW, Suite 300, Boca Raton, FL 33487-2742

and by CRC Press
2 Park Square, Milton Park, Abingdon, Oxon, OX14 4RN

© 2022 selection and editorial matter, Ravi Tomar & Sarishma Dangi;
individual chapters, the contributors

CRC Press is an imprint of Taylor & Francis Group, LLC

Library of Congress Cataloging-in-Publication Data
Names: Tomar, Ravi, author. | Dangi, Sarishma, author. Title: JavaScript : syntax and practices / Ravi Tomar, Sarishma Dangi.
Description: First edition. | Boca Raton : Chapman & Hall/CRC Press, 2022. | Includes bibliographical references and index. | Summary: "Javascript is the most prominent web programming language in the industry today and has endless capabilities in full stack web applications. JavaScript is most preferred because of its compatibility with all the major browsers and its flexibility with the syntax it holds. Being a Front-end language, JavaScript is also used on the server-side through Node.js. This book focuses on developing the basics concepts of Javascript and enlightening the readers about the horizons that can be accessed using this language. Offers detailed explanation of the core topics Covers both miscellaneous and advanced topics Gives a platform to connect JavaScript to cutting edge technologies such as Cloud, Machine Learning, Internet of Things etc Provides examples to enable ease of learning Includes exercises to get more comfortable with complex code Uses case complete projects with examples This book is primarily aimed at undergraduates and graduates studying web application development. Developers will also find it useful as a handbook"– Provided by publisher.
Identifiers: LCCN 2021032161 (print) | LCCN 2021032162 (ebook) | ISBN 9780367641429 (hardback) | ISBN 9780367641474 (paperback) | ISBN 9781003122364 (ebook) Subjects: LCSH: JavaScript (Computer program language) | Object-oriented programming (Computer science) | Programming languages (Electronic computers)--Syntax. | Web site development.
Classification: LCC QA76.73.J39 T64 2022 (print) | LCC QA76.73.J39 (ebook) | DDC 005.2/762–dc23
LC record available at https://lccn.loc.gov/2021032161
LC ebook record available at https://lccn.loc.gov/2021032162

ISBN: 978-0-367-64142-9 (hbk)
ISBN: 978-0-367-64147-4 (pbk)
ISBN: 978-1-003-12236-4 (ebk)

Access the companion website: https://www.routledge.com/9780367641429

DOI: 10.1201/9781003122364

Typeset in Palatino
by MPS Limited, Dehradun

To my loving parents, my wonderful partner Manu and my two little princesses Kritika & Vedika. The continued support from my family and friends could make this book possible.

- Ravi Tomar

To my parents, Rajender and Sushila.

I am grateful for your nurturing and contribution which bought me here. I promise you that I will never give up and continue to strive with my brightest light. I will leave a trace in this world, one with your names behind it. Thank you for all the love, light, support, warmth and especially the wings you gave me to fly.

- Sarishma Dangi

Contents

Preface

JavaScript emerged as a quick fix language, but in the last decade, it has embedded and closely integrated itself with the web. It is difficult to avoid JavaScript while working with dynamic web content. Due to its prevalence, it was soon accepted as a general-purpose language for working in both client- and server-side programming. The popularity, ease of use and rapid adoption of JavaScript made imperative for us to discuss it for novice as well as experienced users. This book aims at covering the JavaScript language basics and fundamentals to help readers strengthen their programming. It is designed to explain the language in the simplest possible words with a comprehensive hands-on approach toward learning by practice. Every chapter in the book is supplemented with demo code, which will compel you to practically test what you read in the chapters. These live exercises allow easier learning for readers by switching swiftly between theory and practice. Lastly, we provide extended case studies that aim to assist readers in exploring more dimensions of JavaScript as a language. There is so much you can cover in a book, right? With that said, we hope that this book offers you an amazing experience at hands-on learning, and you can come back to it whenever you need a little refreshment.

Ravi Tomar
Sarishma Dangi

Acknowledgements

Writing a book resembles a marathon. It is a long and tedious process. This work would not have been the way it is if not for the support provided by two very special individuals.

A special thanks to Dibyasom Puhan, Undergraduate student at School of Computer Science, University of Petroleum and Energy Studies, Dehradun for his untiring support in developing hands-on code and flawless execution of work. Dibyasom is always a dedicated and sincere individual to trust upon.

Other big thanks go to Kamakshi Negi for her contribution and support throughout the writing of this book. Many times, it is hard to move on. In those moments, she gave us light and never hesitated to step in and contribute to this work. We are humbled to have Dibyasom and Kamakshi with us. We wish them joy, success, and warmth wherever they go in the future. Needless to say, we are always a phone call away.

Authors

Dr Ravi Tomar is currently working as Associate Professor in the School of Computer Science at the University of Petroleum & Energy Studies, Dehradun, India. He is an experienced academician with a demonstrated history of working in the higher education industry. Skilled in Programming, Computer Networking, Stream Processing, Python, Oracle Database, C++, Core Java, J2EE, RPA and CorDApp. His research interests include Wireless Sensor Networks, Image Processing, Data Mining and Warehousing, Computer Networks, Big Data Technologies and VANET. He has authored 60+ papers in different research areas, filled four Indian patent, edited 5 books, and have authored 4 books. He has delivered Training to corporations nationally and internationally on Confluent Apache Kafka, Stream Processing, RPA, CordaApp, J2EE and IoT to clients like KeyBank, Accenture, Union Bank of Philippines, Ernst and Young and Deloitte. Dr Tomar is officially recognized as Instructor for Confluent and CordApp. He has conducted various international conferences in India, France and Nepal. He has been awarded a Young Researcher Award in Computer Science and Engineering by RedInno, India in 2018, Academic Excellence and Research Excellence Award by UPES in 2021 and conferred by Research Excellence Award 2021 by Hon'ble Chief Minister of Uttarakhand State, India.

Ms Sarishma Dangi is currently working in the capacity of Assistant Professor in the Computer Science and Engineering department at Graphic Era University, Dehradun, India. She is a Google Cloud Certified Professional Cloud Architect specialized in the area of virtualization and cloud computing. She is a bright young enthusiast driven towards innovative research work and dedicated to imparting quality education to the higher education industry. She has authored many research papers in recognized journals and has authored a book on "Mobile Cloud Computing: Principles and Paradigms". Her research interests include blockchain, virtualization, vehicular ad hoc networks and live memory forensics. She is currently pursuing her PhD with full rigour.

1

Introduction

The kind of programming that C provides will probably remain similar absolutely or slowly decline in usage, but relatively, JavaScript or its variants, or XML, will continue to become more central.

—Dennis Ritchie

JavaScript is one of the most popular languages for web development today. It is a simple, easy-to-use and easy-to-learn language that provides a flexible working style. That is why it is now being used in other areas as well such as for executing server-side scripts, for cloud-based services and sometimes for machine learning applications. In this chapter, we introduce the language itself with its features, advantages and limitations. As JavaScript is run using a web browser, learning about it becomes significant. The structure of modern browsers is discussed along with the historical background of the language and its evolution till now. Lastly, we will learn about some of the methods for displaying information to the user and learn about the placement of code. It is advisable for the readers to practice as they go through the text. That is the best way to learn the language. JavaScript is amazing, and you'll enjoy learning it. It can be frustrating and maddening at times when you can't get it to work, but when you do get things to work the way you want them to work, it's super satisfying.

1.1 Introduction to Web Development

Web development is the process of developing a website to serve static or dynamic content to the user. During the early phase of web development, the aim of developers was to deliver static content to users and allow exchange of information across the internet. Slowly, the opportunities of using web for business and real-time user interaction emerged, leading to a new dynamic world of opportunities. Many technologies, programming languages, design frameworks and approaches have emerged since then. However, the foundations remain the same.

The entire World Wide Web is fundamentally based on the triad of three technologies as shown in Figure 1.1, i.e. HTML (Hypertext Markup Language) [1], CSS (Cascading Style Sheets) [2] and JS (JavaScript) [3]. HTML is a markup

DOI: 10.1201/9781003122364-1

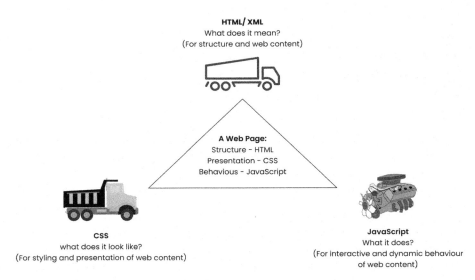

FIGURE 1.1
Triad of web technologies.

language used to structure, tag and put content on the web pages. The primary aim of HTML is to provide information to the users of the web page, and it uses its tags to mark and structure that information. As the web evolved, a need for designing and beautification of web pages soon emerged. CSS was thus integrated with HTML for making the web pages more pleasing to the users' eyes. It greatly helps in managing the presentation and layout of the web pages to make them user-friendly as well as attractive. Finally, JavaScript emerged to give a dynamic behavior to the web pages. This dynamic behavior led to interactive web pages that allowed the web developer to interact with users locally on from the web page itself.

Combination of HTML with CSS made web content more informative and beautiful in terms of presentation. The next stage of evolution was the need of embedding behavior within these web pages to make them more alive and thus *dynamic* in nature. JavaScript is used for altering or defining the behavior of the web pages and is thus referred as the *web enhancing technology*. It dictates how a web page will perform or act to different sets of inputs and behaviors exhibited by the user; for example, activation of a link, checking user input before submitting and link expansion. Combining these three technologies unleashes the full power of web development. Initially, JavaScript was used only for client side, but today it has extended its usage even on server-side technologies [4].

1.2 Client-Side and Server-Side JavaScript

To understand the difference between the client side and the server side, we will use this very simple example.

Suppose we have a very basic website that asks for a number from a user and can print the corresponding mathematical table of that number. Now the development of this application will have following components:

a. A static input page hosted on remote webserver (server side).
b. A processing application server with webserver (server side).
c. A client user using any browser to connect to the webserver (client side).

The working of this application can be broken down into following steps:

a. The webserver is and hosting the static input page; this page can be directly accessed using domain name or IP address.
b. The client connected to internet opens up the web browser and enters the domain name or IP address.
c. The webserver serves the input page to the client.
d. The webpage contains all HTML + CSS code within it; this code is parsed by the web browser and displayed to client.
e. The client now enters the number for example 2 and sends this request back to the webserver.
f. This request will be parsed by the webserver and passed on to the application server. The application server will now generate the web page on the go for this request and send it back to the webserver.
g. The webserver will send this dynamic web page back to the user.
h. The response still contains all HTML + CSS code, but this time it is generated dynamically based upon the input provided by the user.

In this example, the steps performed in (e)–(h) can be omitted with the use of *client-side JavaScript*. The webserver can have JavaScript code inside the static input page itself (HTML + CSS + JS); on receiving this page, the web browser can execute JS code on local machine, and generation of dynamic page can be done locally on the client side only.

This is a very basic example to understand how the number of request/ response calls over network can be reduced. You will find the use of client-side JS in similar situations like getting data from the remote server, using JavaScript to visualize the data, establish asynchronous database connection and many more such cases.

The latest technologies and framework like Node.js [5], React.JS [6] and Angular.JS [7] help in deployment of advance backend application server. All these frameworks will use server-side scripts.

The web browser along with the system running it and the user all together form the *client side* of the application. The *server side* of the application comprises the webserver, application server and the associated applications. Web operates dynamically because of interaction between these two parties. The Figure 1.2 depicts the flow of communcation between Client and Server.

In the early phases of web development, the user requested a page from the server, and the server, on accepting the request, searched the database and responded with the marked page. HTML was used as a markup language to mark the sections within the page containing information. Once the page was sent, the server had no control until the user responded with another request.

With the increase in processing capacity of the user's browser as well the server, the web emerged with a new type of web-based applications. A server responded to a user's request and sent a *web application* instead of a static page, called as the *client-side application,* which executed on the user's browser. These were small applications embedded within the HTML code of a web page and executed dynamically to the user's responses with no need of going back and forth to the webserver.

1.3 Origin, History and Evolution of JavaScript

JavaScript was created by Brendan Eich in 1995 with the goal of making *web* interactive via the NetScape Navigator web browser [8]. In 1995, World Wide Web was gaining popularity, and Netscape Navigator was one of the most popular web browsers at that time. HTML was primarily used to create static content that was displayed on the web pages. With the introduction of JavaScript, those web pages were made responsive as well as interactive. This added to the overall value and end-user experience of exploring the web.

JavaScript was initially developed under the code name *Mocha*. It was named *LiveScript* when it was released in *beta* in September 1995 [9]. It was again renamed in December 1995 as *JavaScript*. One misconception about JavaScript is that it is related to Java programming language, but Java and JavaScript have very less in common except the little similarities in syntax designing. Java was a very popular language at that time, and using the name JavaScript by Sun (now Oracle) was a marketing strategy. In the words of Christian Heilmann, *"Java is to JavaScript what Car is to Carpet"* [10].

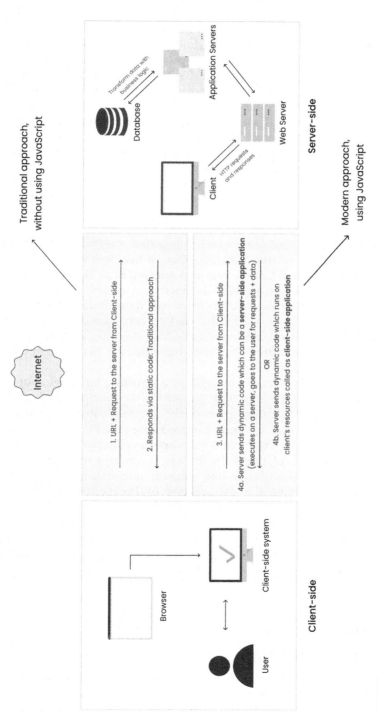

FIGURE 1.2
Client- and server-side scripting.

JavaScript became famous in the web market, and by 1996, it was formally sent for standardization to ECMA (European Computer Manufacturers Association) [11]. The standardization process took some time and was officially called as ECMAScript or ECMA-262 [12]. However, because of its popularity, the name JavaScript was more adopted by people in comparison to the official name. ECMA handles the development and standardization of JavaScript till this day. It ensures that newer versions are compatible, thereby allowing interoperability across web browsers. Many versions of the language have been released since 1996. JavaScript 2 and JavaScript 3 were launched in 1998 and 1999, respectively.

Use of JavaScript by developers kept on advancing, and a breakthrough came in the year 2005. James Garett released a paper describing a set of technologies that were supported by JavaScript. He created the term *AJAX* in the same context. Moreover, JQuery [13], Dojo [14] and Prototype were also released in the same year, which integrated well with JavaScript. All these factors further enhanced the adoption of JavaScript.

ECMAScript Technical Committee, also referred to as TC39, is responsible for maintaining the language [15]. JavaScript standards are now revised annually by TC39, and the modern browsers then update themselves according to the specification. This has led to continuous development of JavaScript and has solved many issues, which it earlier had. All the collaborative efforts from different fronts have led to wide acceptance of JavaScript in the market.

In the beginning, JavaScript emerged as a scripting language, but all the aforementioned developments with support of JavaScript standard development led to a total transformation of the language. It evolved into a robust, secure, dynamic and a general-purpose language that is well suited for front-end, back-end and large-scale software development.

Today, around 90% of the websites use JavaScript in their code. This huge adoption and popularity has made it important for the new developers to learn about the capabilities of JavaScript. New features, technologies, frameworks, plug-ins etc. are constantly being integrated into JavaScript. It has evolved to become a de facto standard web programming language. There are many unique features that have made JavaScript a widely adopted language. These features are discussed in the next section.

1.4 Features of JavaScript

Majority of web pages use JavaScript to make the web more dynamic and responsive to users. JavaScript is a high-level, dynamic, interpreted, prototype-based programming language with support for object-oriented, event-driven and functional programming styles. In this section, we highlight some of the key features of JavaScript that made it so popular among developers.

- **Platform independent:** JavaScript is a platform-independent programming language. Thus, it can be run on any platform making it a versatile and easy-to-use language. It is supported by most of the browsers that have built-in interpreters for JavaScript.

- **DOM manipulation:** DOM stands for Document Object Model that enables access to HTML components within a web page for manipulation [16]. During the early days of web development, most of the web was static in operation. DOM manipulation was one of the key reasons for birth of JavaScript, and its introduction has changed the face of modern web.

- **Date and time:** JavaScript provides in-built methods for operations that need access to date and time utilities. This allows the user to have customized content on their web browsers.

- **Client- and server-side programming language:** JavaScript initially provided client-side scripting only. Today, JavaScript provides the ability to run both client-side and server-side applications. This has made it even more popular as a developer can develop complete applications only by using JavaScript.

- **User data validation and client-side data processing:** JavaScript is used to validate the user input data on the client side and omitting the need to send to server. JavaScript provides a greater control to the browsers than the web servers as the code is executed on the client side. It also allows the webserver to check the browser type, OS, screen size etc. All of this helps to generate a more meaningful script with a less overhead on the server side and more utilization of resources on the client side.

- **Using functions as objects:** Functions in JavaScript are treated as objects that can have both properties and methods. Functions can be passed as arguments to other functions, can be nested, can be thrown as exceptions etc. This versatility to play with functions makes the language easy to operate and more suited to tackle real-world-use cases.

- **Arrow functions:** These are included in the language to make the syntax simple to use and write. They are lightweight functions and can be used to write anonymous functions.

- **Template literal:** It allows the developer to store variable values into strings and thus serves as an important tool. This allows the developer to focus more on the logic and development of application and less on the syntax nitty-gritty.

- **Event handling:** Whenever an event happens, a corresponding routine for that event occurrence can be executed. This feature is called event handling, and JavaScript uses it heavily to respond to user activities on the web page such as *onClick, onLoad, onSelect* behavior.

1.5 Advantages and Limitations

Like every other programming language, there are inherent positives and negatives associated with using JavaScript as well. In this section, we outline the advantages and limitations of using JavaScript. Following are the advantages that JavaScript provides:

- **Speed of execution:** JavaScript is an interpreted language, meaning that the code does not need to compile before execution. The statements in JavaScript code execute as they load in the browser. Overall, this leads to decrease in overhead and increase in the speed of execution of JavaScript code in web page.
- **Easy of learning and simplicity:** JavaScript is very easy to understand, learn and use by new as well as experienced developers. Use of a simple structure and feasibility to implement dynamic content make JavaScript a popular choice for web development.
- **Interoperability:** JavaScript can be easily embedded in any other programming language or inside a web page. Developers use it easily to add new features to their content without any additional overhead.
- **Client-side execution and validation:** The script is sent by the server to the client, and the client web browser executes the script. The data validation also takes place at the client side mostly instead of querying server side each time any data is entered. This reduces the load on the server side and saves a lot of execution time.
- **Rich interfaces and extended functionality:** JavaScript provides a vast set of rich interfaces to enable developers for creating rich web pages such as drag and drop windows, sliders etc. Use of such interfaces provides an increased level of user interactivity on the web page. JavaScript also provides extended functionality by allowing the use of third-party add-ons. This lets the developer to reuse the code and makes the programming much faster and easier to do.

Although JavaScript provides a considerable number of features and advantages, it also has some limitations, which are listed as follows:

- **Reading/writing to files:** JavaScript does not allow the web pages to read or write content to files on the client-side. The access to Operating System is not permitted by JavaScript, considering the security perspectives in view. However, this is enabled in HTML5 Local Store that is provided by the web browser.

- **Multithreading:** JavaScript does not provide multithreading or multiprocessing capabilities. Web browsers do not allow a single web page to use multiple threads concurrently to avoid concurrency issues. However, multiple tabs can have their own individual threads and work around them.

- **Security on the client-side:** As the code is executed on the client-side, the code becomes vulnerable to modification. The source code of a web page is visible to anyone/everyone on the client-side. Malicious users can tamper with the source code or place some other code into the website. This can lead to compromise in security on the client-side.

- **Browser support:** Different browsers interpret the code differently. Therefore, the code must be run on multiple browsers before publishing it for final use. This creates unnecessary overhead for developers to test and run their code multiple times so that the web page doesn't break when it is loaded.

- **Debugging:** JavaScript is a very flexible, prototype-based language, and thus, it sometimes becomes difficult to identify where the error came from. Inheriting from multiple prototypes, interfaces and objects creates dense relationships. Web browser does not specifically show the error, unless it is explicitly handled. A single error may stop rendering the rest of the web page. This leads to difficulty in debugging of large-scale programs.

- **Inheritance:** JavaScript only supports single inheritance. There is no support for multi-inheritance. This becomes an issue while considering object-oriented aspects in mind.

- **Overhead:** JavaScript stores numbers as 64-bits, while bitwise operators operate on 32-bit numbers. Due to this, JavaScript converts a 64-bit number to a 32-bit equivalent, operates on it and then converts it back. Continuous conversions take more time and thus reduce the speed of execution. This back-forth conversion induces overhead in the program. Another reason for overhead is the slow rendering of JavaScript's DOM, thus leading to additional overhead irrespective of the interpreter's execution speed.

All these factors limit the use of JavaScript in certain cases, but anyhow, the advantages far outweigh the limitations. Thus, it is widely accepted in the market by developers and has shadowed most of the web.

1.6 Structure of Browsers

A web browser is a software application used to display web content to the user. It is a fully functional software that performs a series of tasks such as requesting, retrieving, processing and displaying the web content to the end user. It is a Graphical User Interface used to interpret web-based content, including static content wrapped in HTML, CSS, JavaScript-based interactive code etc.

The system on which a web browser runs is the *client side* in client server model of web development. Plugins can also be installed, which help in extending the capabilities of a browser. There are many types of web browsers available in the market today such as Mozilla Firefox [17], Internet Explorer [18], Safari [19], Chrome [20], Netscape Navigator [21] etc.

A web browser is made up of a set of components that perform a set of specific tasks. These components are briefly explained later and depicted in Figure 1.3.

- **User interface:** It is the space within which the user interacts with the browser and makes request to the web server. Typically, it

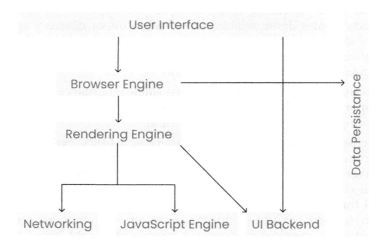

FIGURE 1.3
Structure of browsers.

comprises the address bar, back button, forward button, refresh button, bookmark option, settings, password, history etc.

- **Browser engine:** It is used to take input and execute queries taken from the user interface. The data is then fed to the rendering engine for further processing. Browser engine is the bridge joining the user interface and the rendering engine.

- **Rendering engine:** It is responsible for interpreting and executing the code to display web page to the user. It interprets the HTML, CSS code sent by server as response and displays it on user interface. There are many types of rendering engine, and different browsers use different types of rendering engine. Internet Explorer uses Trident [22], Mozilla Firefox uses Gecko [23], Chrome and Opera use Blink [24] and Safari uses WebKit [25].

- **JavaScript interpreter:** To execute JavaScript code, there is a special component – JavaScript interpreter, which sends its response to the rendering engine. It is then displayed on the user interface.

- **Networking:** Networking component handles the URL retrieval via internet protocols such as HTTP or FTP. It is mainly responsible for maintaining communication and security of the code. It can also implement the cache data to reduce the network congestion.

It is used to creating basic widgets on the browser User Interface (UI) including boxes, widgets etc. It uses the operating system–specific UI methods underneath for its operation.

- **Data storage and persistence:** This component maintains persistence across the network, browser and source code. It is a database present in the storage where browser is installed. Using this storage, browsers are able to support localStorage, IndexedDB, FileSystem and WebSQL as storage mechanisms. Storage component also stores cache, user data, bookmarks, preferences and cookies.

1.7 Saying *Hello World* to JS

Till now, we have discussed some of the concepts used in today's web. It is now time to say *Hello World* to the user using JavaScript. We will also discuss some of the other frequently used methods.

For testing these examples, you can either write complete script provided in the text box in a file saved with *.html* extension and open it in Google Chrome browser or simply open Google Chrome browser and press *Option +* ⌘ *+ J* (on MacOS) or *Shift + CTRL + J* (on Windows/Linux). This will open the console on the web browser. Type the script and below-mentioned code in the console to see the output. Don't worry; we have lots of examples for you to try later with all the details.

JavaScript code is always written within the *<script>* tag. This tag represents the script part of the code and is placed within the HTML document (inside body or head tag), or it is placed in a separate document. This example uses *console.log* and *alert()* method; both are discussed in upcoming section.

```
<script>
    console.log("Hello World on console!");
    alert("Hi This is an Alert for you, you will see this in
pop-up. Have fun!");
</script>
```

1.7.1 Built-In Functions

JavaScript provides some built-in functions to display messages to the users using the browser window. These messages can be displayed via pop-up windows using *alert()*, *prompt()*, *confirm()*, *console.log()* and *document.write()* functions. These dialog boxes provide us with the ability to display information, take input or to get any confirmation from the user. Using these functions, we can interact with the user. Each function has its own way of displaying the message to the user.

- *alert()* is used to display a message on the web page. Using *alert()* triggers a pop-up window with the specified message. It provides a dialog box with a button that displays "ok". This is generally used to provide some warning or information to the user [26].

```
<script>
    alert("Hi This is an Alert for you, you will see this in a
new pop-up");
</script>
```

- *confirm()* is the built-in function used to ask confirmation from the user. It displays the message and provides the user with two buttons, "ok" and "cancel" [27].

```
<script>
    confirm("Are you Sure?");
</script>
```

- *prompt()* dialog box is used to take input from the user. *prompt()* function displays two pieces of information; one is the message or label to be displayed in the dialog box, and the other is the default string to be displayed. It also provides the user with two buttons, *ok* and *cancel*. If the user types in some value and selects *ok*, then the value is returned to the code. If the user selects *cancel*, then the function returns *null* to the calling code [28].

```
<script>
   age = prompt("Enter Age");
   alert("Your entered age value is:" + age)
</script>
```

- *document.write()* is used whenever we want to print the content onto the HTML page. It is a simple, easy-to-use method for displaying content on to a web page [29].
- *console.log()* simply displays the message on the web browser's window. It is primarily used for debugging purposes [30].

```
<script>
   console.log("Hello World via the console!");
</script>
```

1.7.2 Gathering Software

JavaScript code is mainly executed within the web browsers. It is an interpreted language offering faster execution times. Event-based feature of JavaScript makes it more interactive and executed only that segment of code that is required by the user. Modern web browsers all offer support for running JavaScript-based code.

JavaScript ecosystem comprises many tools that help to develop, debug and support developers. There are various tools, frameworks and libraries that are prominently used by JavaScript developers. A very thin line exists between these three, and the distinction can be tricky in itself. Therefore, we define these terms and distinguish them before discussing such tools in detail.

- **Framework:** It is an abstraction that provides a standard way to deliver prominent functionalities, such as events, storage, data binding and so forth. Framework is the most comprehensive means to deliver a set of functionality to the user in the form of well-developed, architected solutions.
- **Library:** It is a collection of functions aimed at addressing certain functionality to perform tasks. In other words, libraries are closely connected set of functions covering a single topic or functionality.
- **Tools:** These are small snippets of code aimed at making development process smoother by assisting developers in maintenance of code, performance, compilers etc.

1.7.2.1 Examples of Framework, Library and Tools

- **React.JS:** It is a library developed by Facebook in 2011 and was later declared as open source, thereby allowing developers to access, use and modify it as they desire. It offers a means to render web content and create dynamic user interfaces with high performance. Components within the library are stand-alone and thus can be reused. Reusing components is helpful while doing system upgrades as it saves a lot of time [6]. React.JS has a wide set of tools that can be used for data binding, virtual DOM and other development. React Native is a mobile application development framework based on React.JS library.

- **Express.js:** It is a JavaScript framework used to build websites with Node.js. Express.js is a robust, stable, fast and easy-to-use framework. It is useful for building single-page applications, HTTP APIs and websites and for routing [31].

- **Angular.js:** It is a JavaScript framework powered by Google, used for building client-side applications. Angular.js is a Model-View-Controller (MVC) type of framework, thereby freeing developers to work on databases, interfaces and linking, as the framework takes care of that. It is a powerful solution in itself to write single-page swift applications. It is a part of the popular MongoDB, Express, Angular, and Node (MEAN) stack for software development [7]. It offers features like cross-platform support, code splitting, command line tools, unit testing and two-way data binding.

- **Electron:** It is a framework used to build cross-platform websites and applications. It is based on Chromium engine and Node.js. It is primarily used for desktop-based applications as it supports HTML and CSS along with JS [32].

- **NPM:** It is a powerhouse of tools offering support for JavaScript developers. It offers to build tools, works as a package and task manager and serves as a registry for offering reusable scripts to developers. Node Package Manager scripts are supported for cross platforms and used for default as well as developer-specific tasks. It offers the ability to design, write and execute independent scripts for testing and development [33].

- **Gulp:** It is a toolkit used for automating tasks related to loading, building, compiling, validating and testing of source files. These tasks are mundane and often take a lot of effort from the developer's side. By using task-specific tools like Gulp, these redundant tasks can be easily automated leading to better performance and code usage [34].

- **Webpack:** It is a tool used for bundling JS modules together along with their dependencies into a browser in an orderly manner. It can split the code for faster performance, optimize at runtime, bundle together tasks, resolve dependencies during compilation and support plugins [35].

- **ESLint:** The process of analyzing the code for the presence of flaws such as a missing colon or bracket is called *linting*. ESLint is a JavaScript linting tool. It supports plugins and can be configured by developers for their use.
- **Vue.js:** It is an open-source, lightweight framework built by using Angular framework aimed at developing progressive user interfaces [36]. It provides tools for routing, dynamic animations and data binding.
- **jQuery:** This is the most popular library for building client-side applications, plugins and user interfaces. Over 70% of websites on internet are built using jQuery including Google, DailyMotion and Microsoft Network [37]. It offers lightweight features, animations, Ajax calls and optimized features.

1.8 Placement of Code

HTML is a markup language that uses tags to place and structure the content. Start and end tags in HTML have a same name. The only exception is that the end tag has a forward slash (/) attached to it. JavaScript code is mostly placed within the HTML tags. JavaScript code is placed within *<script> ... </script>* tags. This can be done by using any of the following ways:

- Inside head tag.

```
<html>

<head>
    <script type="text/javascript">
        //javascript code can be placed here
    </script>
</head>

<body>
    <h1> Test Data </h1>
</body>

</html>
```

- Inside body tag.

```
<html>

<head>

</head>

<body>
   <script type="text/javascript">
                         //javascript code can be placed here
   </script>
   <h1> Test Data </h1>
</body>

</html>
```

- Placing JavaScript code at the end of *<body>* tags is considered a good practice. This ensures that *<head>* and most of HTML *<body>* content are loaded before JavaScript code is triggered.

```
<html>
<head>
</head>
<body>
   <h1> Test Data </h1>
</body>
</html>
<script type="text/javascript">
//javascript code can be placed here
</script>
```

- The JS code can also be kept in an external file and then linking it in *<head>* section of HTML. External file should contain JavaScript code and must be saved with '*.js*' extension. Use of external files for placing JavaScript code is preferred as it helps in separation of the content and behavior of the web page. Maintaining a single source file is easier and also allows us to reuse the same code.

```
<html>
<head>
   <script type="text/javascript" src="filename.js">
</script>
</head>
```

```
<body>
   <h1> Test Data </h1>
</body>
</html>
```

1.9 Exercise

1.9.1 Theory

i. List out the features of JavaScript language.

ii. What are the benefits and challenges of using JavaScript?

iii. Discuss the evolution of JavaScript as a language since its inception.

iv. Elaborate different components present in the modern web browsers. Explain the working of web browsers using different components.

v. Differentiate between client-side and server-side JavaScript usage. Discuss the cases where these must be used. Think and design your own use case.

vi. Write a complete script that prints "Hello.world" using the console.log() and document.write().

vii. Create one HTML document that displays an alert dialog box to the user. Once the user clicks a form button, immediately another alert dialog box must appear on the same page.

viii. While coding for aforementioned questions, carefully monitor the browser display page. How can users interact with the page? What role does script play, and how can you change it?

1.9.2 True/False

i. JavaScript is derived from the Java language.

ii. JavaScript works independently to deliver dynamic web pages to the end user.

iii. JavaScript can be used for client-side programming or for server-side programming or for both.

iv. JavaScript code can be written in an independent file and then embedded within HTML document.

v. JavaScript supports multithreading.

1.9.3 Multiple-Choice Questions

    i. JavaScript code is written inside the _____ tags.
      a. \<html\>
      b. \<head\>
      c. \<script\>
      d. \<body\>

    ii. Which of the following languages make the triad for web technologies?
      a. HTML, XML, CSS
      b. CSS, HTML, JavaScript
      c. JavaScript and XML
      d. None of the above

    iii. Which of the following is not a feature of JavaScript?
      a. It is lightweight interpreted language.
      b. It is a prototype-based, object-oriented language.
      c. It is a markup language.
      d. It can handle date- and time-related events.

    iv. JavaScript was developed by _____ company.
      a. Dell
      b. Bell Labs
      c. Netscape
      d. Microsoft

    v. What was the original name for JavaScript, back when it was conceived?
      a. LiveScript
      b. NewScript
      c. Mocha
      d. JavaScript

    vi. JavaScript is a _____, _____ programming language.
      a. high-level, object-oriented
      b. object-oriented, prototype-based
      c. high-level, assembly-based
      d. None of the above

2

Building the Basics

You should imagine variables as tentacles, rather than boxes. They do not contain values; they grasp them—two variables can refer to the same value.

–Marijn Haverbeke

Till now, we have covered the history and introduction of JavaScript. This chapter will take you to a deep dive in understanding the language constructs and fundamentals. To build solutions using any language, we need to have an understanding about how the data is stored and transformed. This chapter discusses fundamental building blocks such as learning how to declare variables, using data types and manipulating them, learning about the lexical structure of a document, various operators for data manipulation, the decision statements and control flow operations.

2.1 Lexical Structure

Lexical structure defines the lowest level of tokens for any programming language. Lexical structure is the set of elementary rules which are followed to form valid tokens from character sequence [38]. A token consists of an identifier, keyword, punctuation, literal or operator. Language specifics such as character set, variable naming, comments, terminators etc. all are used to define the lexical structure; for example, it defines how a variable should be named, how you can incorporate comments in your code or put line breaks etc.

2.2 Character Set

A character set defines the valid characters that can be used in source programs or interpreted when a program is running [39]. The JavaScript uses the 16-bit representation of the Unicode character set for programming; this allows the programmer to use certain symbols as identifiers in any language that are supported by UNICODE like Π (pi) or λ (lambda). Unicode character set is a superset of the ASCII character set [40]. For code portability and ease of use, mostly programmers use ASCII character set in their code.

DOI: 10.1201/9781003122364-2

2.2.1 Whitespace and Comments

Whitespaces are used to separate different tokens present in the code, and the JavaScript interpreter ignores whitespaces found within the code segments or between tokens in the program. This allows us to use line breaks and whitespaces intentionally to indent our code and make it easier to understand. It helps in making the code neat and consistent for use. JavaScript allows us to use newlines, carriage returns and line feed sequences as line terminators. Line terminators can also be used to create a definitive style for writing your code.

The following characters are considered whitespace: space (U + 0020), line feed (U + 000A), carriage return (U + 000D), horizontal tab (U + 0009), vertical tab (U + 000B), form feed (U + 000C) and null (U + 0000).

Many a times, it is possible that a code written by one developer is debugged, maintained or handled by another developer in future. Proper use of comments helps the new developer to understand and work with the code. There are systems that can even generate a meaningful report from the comments available within the code. Comments are also treated as whitespace by the interpreter when the line begins with // and ends with a line feed. Multiline comment is denoted by using the symbols "/*" and "*/". Anything that falls in between these symbols is considered as a comment. Usage of both types of comments is demonstrated in the following example.

```
//Demo Comments
<script>
  document.write("Hey there, reader")
  // printing a line of code
  /* THIS IS A MULTI LINE COMMENT
  THIS LINE WILL ALSO BE OMMITED */
</script>
```

2.2.2 Case Sensitivity

JavaScript is a case-sensitive language, i.e. *name*, *Name*, *NAME* and *NaMe* are all different from one another. While writing your code, be aware of the capitalization of letters. Mostly, the code is written in lowercase letters because it is easy to read and avoids unnecessary complexity. Standard coding practice is advised for variable and function naming. *firstNumber* and *addNumbers()* are examples of camel casing convention for naming a variable and function, respectively.

2.2.3 Semicolons

Most languages like C and Java use semicolons to terminate a line of code. Unlikely, JavaScript has optional requirement of semicolon and thus allows flexibility to the developer. However, if you are writing multiple statements in a single line, then semicolons are used to tell the interpreter that there are multiple statements in the line, which need execution separately.

2.2.4 Literals

Literal is simply a source code representation of a value of a type. The value of literal is used as is in the program and does not have a type. The literals can be numeric, decimal, string and Boolean. The example is given next.

```
89                  //Integer Literal
7.99                //Floating Point Literal
"Hello JavaScript"  //String Literal
true                //Boolean Literal
```

2.2.5 Identifiers

Identifiers are the names given to the literals, variables, function, property or a class. The name is used to uniquely identify a variable, function, property or a class. Identifiers are unique throughout the scope, and following are strict naming conventions that must be followed while naming an identifier [41].

- Identifier name must start with a letter (a–z or A–Z), underscore (_) or a dollar ($) sign.
- After first letter (a–z, A–Z, _, $), we can use digits (0–9) also.
- Identifier names are case sensitive.
- Reserved keywords cannot be used as identifier.

Based upon the aforementioned constraints, some of the valid identifiers are given as:

```
name;       //valid name
Name;       //valid "name" and "Name" are different
_id;        //valid name
#id;        //not allowed
_revisedID  // allowed
void;       //not allowed, 'void' is a reserved keyword
```

2.2.6 Keywords

Every programming language reserves some keywords that are used as a part of the language itself [42]. These keywords are also called reserved keywords. They cannot be used as identifiers. Following are some of the reserved keywords for JavaScript language; the list is not limited to these (Table 2.1).

TABLE 2.1

Reserved Keywords in JavaScript

break	*switch*	*do*	*while*
if	*else*	*for*	*forEach*
instanceof	*typeof*	*let*	*export*
as	*finally*	*const*	*package*
var	*return*	*class*	*null*
throw	*void*	*extends*	*new*
catch	*continue*	*function*	*switch*
ifelse	*break*	*yield*	*case*

2.3 Variables

Variables are the names given to storage locations in JavaScript. These storage locations can then be used to store any type of data by directly referring them via their assigned names. The variables names are provided using identifier naming conventions discussed in the previous section. Unlike conventional programming languages (C, Java etc.), JavaScript is a loosely typed language, and there is no need to provide type of the variables while declaring them [43].

2.3.1 Variable Declaration and Scope

The scope of a variable in JavaScript can be either local or global. Based upon the requirement of the scope of the variable, it can be declared by using either *var* or *let* keyword before the valid variable identifier name.

Scope of a variable determines the accessibility of data items within a program. Variables declared within a function are mostly required to be in local scope, as the value is required locally within the function. To create local variables, the *let* keyword is used before a valid identifier; for example:

```
let localVariable;
```

These variables are created when a function or a block starts execution and are deleted once the block scope ends. Variables with same name can be used outside the local scope of a function, i.e. in global scope.

Similarly, if the requirement of a variable is in the global scope, then the variable is declared using a *var* keyword. The global variables can be accessed from anywhere in the program. The example of creation of a global variable is shown next.

```
var globalVariable;
```

If the value is not provided during the creation of a variable, it takes the value of *undefined* automatically until you assign a value to it. Both local and global variables can be initialized during the creation; this is demonstrated in the example next:

```
let localVariable = 10;   //Declare and Initialize
let globalvariable = 20; //Declare and Initialize
```

There are few more differences between the variables declared using *var* and *let* such as the following:

- The variable can be redeclared if created using *var* keyword, while it is not possible with the variables created using the *let* keyword.
- We can use *var* keyword to create global or local variable by defining it outside or inside the block. This is not possible in case of *let* declared variables as they are always local within a block and cannot be accessed using other objects.

These two scenarios can be seen in the next example:

```
var dogName = "American Aketa";   //global in scope

  function dogType() {

    var dogName = "Ojo";
// dogName is redeclared and is also a local variable
  }.
```

You might be landing with so many questions about accessibility of variables: Which variable to call when? Who will decide which value to use and when to use? The answer to all these questions is found in scope of variables!

The scope of a variable defines the set of rules to resolve which variable value to be used and where is variable accessible.

Let us try to understand this example taken from [44] with the help of Figure 2.1. Let us first find the identifiers within areas marked as 1, 2 and 3. The block 1 contains the identifier *foo* (method calling and method declaration); further, block 2 contains the identifiers *a*, *b* and *bar*. The innermost block 3 contains the identifier *c* only. The method *foo* is called with the value of 2 passed to *a* that initializes identifier *b* with 4 inside block 2 and calls other method *bar* with value of 12 passed to the identifier *c* in block 3. This whole path is carried out in the following way, and hence, when the innermost *console.log* needs *a, b, c* to print their values, it will find in the immediate scope and will move in stack order to find appropriate value. The values printed on console will be the one found inside the block 3. The lookup stops as soon as the first matching identifier is found. This approach is called *lexical scope*. The values of variables a,b and c are elaborated in Figure 2.2 within the scope.

```
function foo(a) {

        var b = a * 2;

        function bar (c) {
                console.log( a, b, c );
        }

        bar(b * 3);

}
foo( 2 );                                        // 2, 4, 12
```

FIGURE 2.1
Understanding the scope of variables.

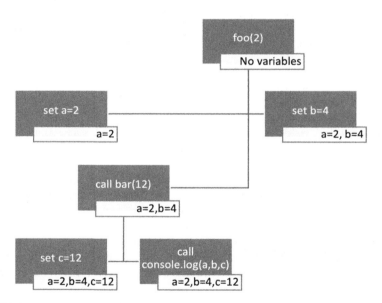

FIGURE 2.2
Elaborating the scope of variables.

2.4 Data Types

Variables in JavaScript can hold any type of data without specifying a strict data type. Such programming languages are referred to as dynamically typed languages or loosely typed languages, i.e. data types exist but are not strictly bound to the variables. There are two basic categories of data types in JavaScript: primitive and nonprimitive data types.

2.4.1 Primitive Data Types

Primitive data types can only store a single value within the associated variable. It cannot store additional properties or methods along with the data. There are seven primitive data types: *string, number, bigint, boolean, undefined, symbol* and *null*[45]. Except the *undefined* and *null*, all primitives also have a wrapper object in JavaScript. All primitives are by default immutable in JavaScript. This means that we cannot change any primitive value; however, a newly created value can be initialized to other variable. For example, in this code snippet, the variable *someString* is not mutated until the reassignment is done. This indicates that a new "DUMMY STRING" is created and assigned to the *someString* identifier.

```
// Using a string method doesn't mutate the string
var someString = "Dummy String";

console.log(someString);        // Dummy String

someString.toUpperCase();

console.log(bar);               // Dummy String

// Assignment gives the primitive a new (not a mutated) value

someString = someString.toUpperCase(); // DUMMY STRING
```

The following section explains the primitives and their corresponding wrapper objects; the *valueOf()* function can be used to convert any object to its corresponding primitive data type.

- **String:** A string is a sequence of characters representing text value. A String data type object in JavaScript can hold character or string values. The values must be enclosed within quotes (double quotes, single quotes, backticks). Double quotes ("...") and single quotes ('...') are similar in functioning, and there is no marked difference. However, backticks (`...`) are used to embed data value in the script [46]. Also known as extended functionally quotes, they allow us to embed either expressions or variables into a string by using ${...} symbol.

```
let str1 = "our first string";     //allowed
let str2 = 'our second string';        //allowed
```

```
let str3 = `we can embed ${str1} and ${str2} in this`;
//embed a variable

/*output: we can embed our first string and our second
string in this*/

let str4 = `embed an expression ${25 + 25}`;
//embed an expression
        //output: embed an expression 50
```

- **Number:** In JavaScript, number is a numeric data type in the double-precision, 64-bit, floating point format (IEEE 754) [47]. A Number data type object in JavaScript can hold any type of numerical value including both integer and floating point numbers. For example:

```
let a = 78              //normal numeric value allowed
let a = 7895.00         //value with decimals is allowed
let a = 456e3       //456000 large values allowed
let a = 456e-6      //0.000456 small values allowed
```

Special numeric values are also allowed in JavaScript, i.e. *Infinity, -Infinity* and *NaN* (NotANumber). Infinity represents a special value that is greater than any number. NaN is a value

- **Boolean:** A Boolean is a logical data type; it can have either a true value or a false value. Boolean is mostly used in conditional statements and for other testing operations. The wrapper object for *boolean* data type is the *Boolean* object [48].
- **BigInt:** In JavaScript, *bigint* is a numeric data type that can represent integers in the arbitrary precision format. The wrapper object for *bigint* is *BigInt*[49].
- **Symbols:** A value having the data type Symbol can be referred to as a *Symbol value*. In a JavaScript runtime environment, a symbol value is created by invoking the function Symbol, which dynamically produces an anonymous, unique value [50]. The wrapper object for *symbol* is *Symbol*.
- **Undefined:** Undefined primitive value is assumed by a variable when you declare it but do not assign a value to it. It is a special data type and can be technically assigned to a variable. The *undefined* primitive does not have any wrapper object [51].

```
let y;
alert(y);       //output: undefined
```

```
y = 46757;
alert(y);                //output: 46757
y = undefined;
alert(y);                //output: "undefined"
```

- **Null:** As the name signifies, null represents an *empty* or an *unknown value* to a nonexistent or invalid object or address. In JavaScript, *null* is marked as primitive, but in certain cases, *null* is not primitive (because later we will see that all objects are derived from *null*, and hence, *typeof* operator may return object for *null)*. The *null* primitive does not have any wrapper object [52].

```
typeof null === 'object' // true;
```

2.4.2 Nonprimitive Data Types

The nonprimitive data types in JavaScript are used to store more complex values in a variable. It also allows us to associate different properties and methods along with the data in a variable. Nonprimitive or reference data types can be categorized into three types: Object, Array and RegExp. The Object represents the instance through which we can access the members, Array represents the group of similar values and RegExp represents the regular expression. The nonprimitive data types are so important in JavaScript that these will be discussed at great length in upcoming chapters.

2.4.3 Constants

While programming, we may also come across a situation when a value of an identifier once assigned should not be altered again. The variables and identifiers declared via *let* or *var* keyword are mutable in nature, i.e. their value can be changed to other literal or variable. This is where the concept of constant comes into play. The keyword *const* is used to declare a constant, i.e. their values cannot be changed once assigned. For naming convention, the constant name is always written in capital letters.

```
// const CONSTANT_IDENTIFIER = assigned_value;
const PI = 3.1415;
const ROI = 6;
const PUBLISHER = `T&F`;
ROI = 7;                              //Type Error
```

Always remember that the *const* keyword makes the memory allocation to an identifier to be immutable, i.e. the identifier cannot be reallocated to any other variable or memory. However, the content inside memory location can change; for example, if a constant array or object is created, the content may change, but the constant identifier cannot change to another object or array. Keep this point in mind while you learn objects in further chapters. For a quick reference, see the next example. Here you can see that the content inside object can be altered and even added, but we cannot reassign the constant identifier to new value.

```
const SUBJECT = { sub1:"phy", sub2:"math", sub3:"chem" }
//creating a constant identifier "SUBJECT"

console.log(typeof SUBJECT);      //will print object
SUBJECT.sub3 = "bio";             //mutable property
SUBJECT.sub4 = "english"          //add new property
SUBJECT = { sub1:"accounts", sub2:"economics", sub3:"-
statistics" }
                      // CANNOT reassign a constant object
```

2.4.4 Type Casting

Type casting is the process of converting one data type to another data type. There could be situations when we need this conversion for the program, for example, to print a numeric value in a HTML tag where strings are only expected. To help this conversion, JavaScript provides implicit and explicit conversion alternatives [53]. The implicit conversion enables automatic conversion from one type to another when assigned or used in methods. For example, if we print a numeric value in *console.log()* method, JavaScript will automatically call the *toString()* method over it. Similarly, the use of + operator with strings and numeric values automatically converts the numeric value to a string value.

```
var age = 35;
console.log(typeof age);      // typeof age is number
console.log(age);      //equivalent to console.log
(age.toString());
var output = "My age is"+age;      //Implicit Conversion
console.log(output);      //displays My age is 35
```

If JavaScript is unable to do implicit conversion, then explicit conversion is needed. For example, suppose we have two numbers *numberA* and *numberB* that are of String type, and we need to perform arithmetic addition on them. In this case, we will have to explicitly define the type casting being performed. The next example shows both these cases.

```
var numberA = "10";
var numberB = "20";
console.log(typeof numberA);      //type is String
console.log(typeof numberB);      //type is String
```

```
console.log(numberA + numberB);    //Output is 1020
console.log(Number(numberA) + Number(numberB));
//Output is 30
          //This is explicit type casting
```

JavaScript offers explicit conversion by using built-in methods such as *Number(), String(), Boolean()* etc. Empty strings and using null value in JavaScript always return 0.

2.5 Operators

Mathematical and other operations are performed using operands and operators. Operands are the data items on which operators execute [54]. The operators are the symbols that represent the type of operation to be performed on the operand, i.e. data item to get desired results. Operators help in performing arithmetic, logical, assignment and bitwise operations. The combination of operator and operand is called expression, which evaluates to provide the end result. For example:

```
var a = 5 + 10; //expression comprising operator and operands
```

Here + and = are arithmetic and assignment operators, respectively. a, 5 and 10 are operands on which operators operate. Operators can also be classified based upon the number of operands they work upon; for example, unary, binary and ternary operator. Unary operators work on single operand, binary operand works on two operands and ternary operator operates upon three operands.

Following sections cover different categories of operators based on their type such as arithmetic, logical, bitwise etc. It is advised that a novice programmer must practice with these operators to get a hold on their working and use.

2.5.1 Arithmetic Operators

Arithmetic operators take numeric value (literal or variable) as operand and return a single numerical value [55]. JavaScript offers basic arithmetic operators including unary and binary operators such as + , -, * and /. Table 2.2 provides arithmetic operators in JavaScript along with their descriptions and examples.

2.5.2 Comparison Operators

The comparison operators are used to compare the operands and return a logical output. The operands can be numerical, string, logical or object

TABLE 2.2

Arithmetic Operators

Operator	Description	Example	Value of a
+**Binary**	Addition	let a = 5 + 12	*a = 17*
-**Binary**	Subtraction	a = a-3	*a = 14*
*****Binary**	Multiplication	a = a*2	*a = 28*
/**Binary**	Division	a = a/4	*a = 7*
%**Binary**	Modulus or remainder	a = a%2	*a = 1*
++**Unary**	Increment	a + +	*a = 2*
--**Unary**	Decrement	a--	*a = 1*
--**Unary**	Unary Negation	-a	-1
+**Unary**	Unary Plus	+a	1
*****Binary**	Calculates the base to the exponent power.	a = 2**3	*a = 8*

values. JavaScript tries to perform comparison based on the type of operand, such as lexicographical in case of string using Unicode [56]. During comparison, JavaScript carries out implicit type casting for almost all operators, except === and !== only. These two operators perform strict comparison without type casting any of the operand. Table 2.3 provides descriptions and reference examples for using comparison operators.

TABLE 2.3

Comparison or Relational Operators

Operator	Description	Example: Let a = 7	Result
== **Binary**	Equal to (type is not considered)	a = =7	true
		a = =8	false
		a = ="7"	true
=== **Binary**	Identical to (date type and value is considered)	a = =="7"	false
		a = ==7	true
!= **Binary**	Not equal to	a!=8	true
!== **Binary**	Not identical to (date type and value is considered)	a!=="7"	true false
		a!==7	
> **Binary**	Greater than	a> 7	false
>= **Binary**	Greater than or equal to	a> =7	true
< **Binary**	Less than	a <7	false
<= **Binary**	Less than or equal to	a <=7	true

TABLE 2.4

Bitwise Operators

Operator	Description	Example	Result (Shown in 4 bits)
&Binary	AND	7 & 1	$(0111)_2$ & $(0001)_2 = (0001)_2 = (1)_{10}$
|Binary	OR	7 | 1	$(0111)_2$ | $(0001)_2 = (0111)_2 = (7)_{10}$
~Binary	NOT	~ 7	~$(0111)_2 = (1000)_2 = (8)_{10}$
^Binary	XOR	7 ^ 1	$(0111)_2$ ^ $(0001)_2 = (0110)_2 = (6)_{10}$
«Binary	Left Shift	8 « 1	$(1000)_2$ « Insert one ZERO = $(10000)_2 = (16)_{10}$
»Binary	Right Shift	8 » 1	Insert one Zero »$(1000)_2 = (0100)_2 = (4)_{10}$

2.5.3 Bitwise Operators

Bitwise operators are used to perform bitwise operations on operands in JavaScript. The numeric values are converted into 32-bit numbers, and then, the bitwise operator is applied bit by bit. The result is converted back into numeric form after applying the operator [57]. It includes basic AND, OR, NOT, XOR, Left Shift and Right Shift operators as described in Table 2.4.

2.5.4 Logical Operators

Logical operators are applied on Boolean operands or any expression returning Boolean; the result is also Boolean [58]. Similar to all other, if operands are not Boolean, then the implicit type casting treats any literal as true and operation is performed accordingly. Table 2.5 provides reference examples for using logical operators. Let us assume x = 5 and y = 8 for the following table:

2.5.5 Assignment Operator

The assignment operator (=) is used to assign the value of literal or variable on right side to another variable on the left [54]. There are also compound

TABLE 2.5

Logical Operators

Operator	Description	Example: Let x = 5; Let y = 10	Result
&&Binary	Logical AND	(x <7 && y> 7);	true
		(x> 7 && y> 7);	false
		false && (3 = = 4);'	false
		Camel' && 'Dog';	Dog
||Binary	Logical OR	(x> 7 || y> 7);	true
		(x> 7 || y <7);	false
		false || (3 = = 4);'	false
		Camel' || 'Dog';	Camel
!Binary	Logical NOT	!(x = =y);	true
		(x = =y);	false

TABLE 2.6

Assignment Operators

Operators	Description	Example	Result
=	Assign right to left	x = 5	x contains 5
+=	Add and assign	x + =3	x contains 8
-=	Subtract and assign	x-=3	x contains 5
*=	Multiply and assign	x*=3	x contains 15
/+	Divide and assign	x/=3	x contains 5
%=	Modulus and assign	x%=2	x contains 1

assignment operators that are shorthand for these operators. Table 2.6 lists various assignment operators coupled with other arithmetic operators to give shorthand notation of using them.

2.5.6 Ternary Operator

The ternary operator is a special operator that operates on three operands and hence *ternary* in nature. The logical expression and two values (or expression) from which one will be used as a result based upon the logical expression or condition being passed. It is useful in writing single line conditional statements, which can replace the *if else* code. The if else statement will be discussed in upcoming section [59]. The Ternary operator is mostly used for testing conditions and then either choosing value1 if the condition is true or choosing value2 if the condition resolves to false. The syntax for using ternary operator is given as:

```
var var_name = (condition)? value1: value2
```

The equivalent *if else* statement code for the above ternary operator would be:

```
if (condition) { var_name = value1 }
else { var_name = value2 }
```

The condition is a logical expression that results in either a true or a false value. If the value of logic is true, then *value1* is assigned to the variable name; else the *value2* is assigned to the variable name. For instance, to test the age of a voter, the following condition can be used:

```
let eligibleToVote = (user.age> = 18)?"Yes": "No";
```

2.5.7 Comma Operator

The comma operator is also a binary operator and is used to put multiple statements in a single line of code. It evaluates each expression starting from left and returns the value of the last expression in the statement. It is used as a separator while writing code, especially at places where one statement per line is expected and multiple statements in a line are provided, all separated by using the comma operator [60]. The comma operator is used heavily while writing code such as in *for* loops to declare multiple initialization variables or conditions, in *array* to separate two entries while using objects etc. For example:

```
var x = [0, 1, 2, 3, 4].          //Array example
var i;
for (i = 0, j = 9; i <= 10; i + +, j--)
//usage of, operator in loops
{
    console.log('a[' + i + '][' + j + ']= ' + a[i][j]);
}
```

2.5.8 Typeof Operator

The *typeof* operator returns the data type of the operand on which it is used. This is a unary operator and is used to know the data type that may require explicit type casting. The data type of the operand is returned as a string value [61]. Syntax for using *typeof* operator is given as:

```
typeof operand
typeof (operand)
```

Derived from [62], Table 2.7 gives the return values when *typeof* operator is used on different data types found in JavaScript.

TABLE 2.7

Return Value for Different Data Types

Type	Return Value via *typeof*
Undefined	"undefined"
Null	"object"
Boolean	"boolean"
Number	"number"
String	"string"
Symbol	"symbol"
Function object	"function"
Any other object	"object"

2.5.9 Delete Operator

The *delete* operator is a special operator and is primarily used with objects that we will discuss in upcoming chapters. It is used to delete or remove a property from an object. It is designed to delete both the value and the property related to an object [63]. It returns true if the property is successfully deleted. The syntax for using *delete* operator is given as:

```
delete object['property']
```

2.5.10 Void Operator

void is a unary operator, used to intentionally return *undefined* or to evaluate an expression without returning the evaluated value. Using void operator will always return undefined value as its result [64]. It can be used with functions, normal operators and with browser in-built methods to purposely return *undefined*.

```
void (7 = = '7');    //compares both and returns undefined
void 7 = = '7';
//compares void 7 i.e., undefined with '7', returns false
```

2.6 Control Flow Statements

While designing and writing code for different scenarios, a sequential flow of code is not always desired or suitable under varying circumstances. There comes a need of branching the set of statements based upon some decisions [65]. Sometimes, the requirement to iterate a set of statements also emerges. These situations require special mechanism to control the flow of the code. JavaScript language provides decision statements and loop iterators to fulfil these requirements. The set of statements is referred as the *block* and is created using a set of curly brackets, i.e. *{}*. The blocks have been used in almost all the examples we have seen so far. One block represents a piece of work done using multiple statements and will be used with the control statements. Let us understand decision statements, loops and iterators in detail.

2.6.1 Decision Statements

As previously discussed, a developer may need to execute different decisions for different actions. Decision statements are those statements that are executed only when they satisfy the specified condition. This section discusses how these decision constructs are implemented in JavaScript.

2.6.1.1 If Else

The *if* statement is used to execute a block or expression based upon a logical condition. The block following the condition is executed only when the condition specified resolves to true. If the condition evaluates to false, the statements following the *if* block are not executed and an optional else statement is executed, if provided. In both cases, the rest of the code is executed. The syntax for using *if else* statements is given as follows:

```
if (Logical Expression)    //specifying logical condition
{
    /* this block of code,
    only executes when condition
    i.e. Logical Expression evaluates to true */
}
else                //This is optional, but useful
{
    /* This block of code,
    only executes when condition
    i.e. Logical Expression evaluates to false*/
}
// Rest of the code will always execute, in a sequential
manner
```

The *if* block can be used independently without using the *else* block. However, vice versa is not true. There must be an *if* block for every *else* block. *if* and *else* blocks can be nested within each other where every *else* block is paired to the closest *if* block.

There could be a situation where one has to choose one option out of multiple alternative blocks, i.e. one block of code needs to be executed (from many provided for different conditions) based upon the result of some condition. For such scenarios, *elseif* statement ladder can be used. The following example illustrates the use of *elseif* statement along with *if* and *else* statements.

```
if (rank <10) {
    console.log("Congratulation, your rank is in single
digit");
}
else if (rank <40) {
    console.log("Congratulations, your rank is in range of
10 to 40");
}
else {
    console.log("Unfortunately, you did not get a rank.");
}
```

2.6.1.2 Switch Case Statement

The *switch* statements provide a more readable way of writing *elseif* statements when the conditions are based upon equivalence, i.e. you need to compare two values using == operator. Switch statements are more versatile than *if else* statements. Switch is also used for executing different blocks of code for different statements. It is more powerful than *if else*, which always evaluates to either true or false values.

Using *switch* statement, one can specify statements that evaluate to give different values or expressions. These values can then be used with different cases where each case corresponds to a specific value. Whenever *switch* block is executed, the statement inside *switch* is evaluated. The resultant value is then matched one by one with the provided cases. If the value matches to a case, the corresponding block of code is executed; else, next case value is matched. If no case matches the value, the *default* block of code is executed. Syntax for using *switch case* statement is given as:

```
switch (statement)    //statement gives result as a,b,c or
other value
   {
       case a:                   //for a, this is executed
       //block of statements
       case b:                   //for b, this is executed
       //block of statements
       case c:                   //for c, this is executed
       //block of statements
       default:                  //for any other value
       //block of statements
   console.}
```

The following example uses *switch case* statement to print the current day of the week; you will notice *break* keyword between the case. This is used to bring control out of the block; if *break* is omitted, all the cases will be evaluated and produce output on the

```
var dayOfTheWeek = 1;

switch (dayOfTheWeek) {
   case 0:
     console.log("Sunday");
     break;
   case 1:
     console.log("Monday");
     break;
   case 2:
```

```
    console.log("Tuesday");
    break;
  case 3:
    console.log("Wednesday");
    break;
  case 4:

    console.log("Thursday");
    break;
  case 5:
    console.log("Friday");
    break;
  case 6:
    console.log("Saturday");
    break;
  default:
    console.log("Sorry, you made a mistake.");
}
```

2.6.2 Loops and Iterations

Loo-ps and iterations are also used as a part of maintaining control flow statements. In particular, these deal with the repetition of a block of code using certain conditions. Generally, the set of conditions and initializer are used to define how many times any block of code should be iterated. During iteration, loops also provide the values that can be used to track counts, which can also be used to iterate over the values present in array or collections [66].

2.6.2.1 For Loop

The *for* loop is the simplest of all and very comprehensive to write. The *for* loops take an initial value, termination condition and increment/decrement condition as input. The initial value is used only once for initialization purposes. The termination condition is tested, and the block is executed. Once the block is executed, the increment/decrement condition is evaluated to update the value of counter. The test condition is checked again, and the code is executed if the counter satisfies the condition. This way the loop continues to execute until the counter value fails to meet the criteria or the termination condition is met. The syntax of using *for* loop is given as:

```
for (initial value; termination condition; statement)
{
    //block of statements to execute
}
```

Let us also understand the usage of *for* by one simple example. In this example, the variable *i* is initialized to 1 and set to execute till the value is less than or equal to 10. Now, the block starts executing for the first time as *i* = 1 and is less than 10; after every iteration the *i* ++ + executes making an increment of 1 in value of *i*. Now the value of *i* is 2 and is still less than 10. This process is repeated till the termination condition meets, i.e. *i* becomes 10.

```
for (i = 1; i <= 10; i + +) {
    console.log("value of i is"+i);
}
```

2.6.2.2 While Loop

The *while* loop is another way of writing a loop that helps in executing a block of code under a condition. In a *while* loop, the initializer, termination condition and increment/decrement counters are managed separately. It is used for simple, repeated execution of code for a given conditional statement. *for* loop combines the initialization, termination and increment/decrement conditions in one statement, whereas *while* loops allow the developer to break it up and design the solution in a different way. The normal construct of using a *while* loop is given as:

```
let i = 1;              //Initialization
while (i <= 10)              //Termination Condition
{
   console.log("value of i is"+i); //block of statements
   i + +;                //Increment/Decrement
}
```

2.6.2.3 do-while loop

The *do-while* loop is a specialized form of *while* loop where the code block is executed at least once and then the condition is checked to determine further execution. If the condition resolves to true, code block is executed again.

The following example illustrates the usage of *do-while* loop. We can see that initial value of *i* is 20; this code will still produce the output once because the block is executed once and then the condition is checked. For using *while* and *do-while* loops, always specify an increment or decrement statement within the code block; else the loop will never terminate and will continue execution indefinitely.

```
var i = 20; //Initialization
do {
   Console.log("value of i is " +i);   //block of statements
```

```
    i + +;                      //Increment/Decrement
}
while (i <= 10);                //Termination Condition
```

2.6.2.4 for-in loop

The *for-in* loop is a special kind of loop JavaScript. It is used to loop over the properties of an object. For every key present in an object, the block of code is executed once. The significance of using *for-in* loop will be more understood in upcoming chapters when we learn about objects and its properties. The syntax for using *for-in* loop is provided as follows:

```
for (key in object) {
    //code block to be executed
}
```

2.7 Exercise

2.7.1 Theory

i. What are the different data types used by JavaScript? Elaborate with examples.

ii. Differentiate between local and global variables. Design a script for using these variables and showcase their working.

iii. What are the available operators supported by the JavaScript language? Give examples for special operators such as typeof, comma, delete, void etc.

iv. Write a program that prints the following pattern on the console. Use control flow loops to generate it repeatedly.
```
*
**
***
****
***
**
*
```

v. Write a script which prints all the numbers between 1 and 50. For every number divisible by 5, print "I am a factor of 5", and for every number divisible by 7, print "I am a factor of 7" on the console. If you are able to do this with ease, then try to print "I am a factor of both 5 and 7" for numbers that are divisible by 5 and 7 both.

2.7.2 True/False

i. JavaScript is a case-sensitive language.

ii. Automatic type conversion is supported by JavaScript.

iii. Bitwise operators are used to test conditions on variables.

iv. A block of statements in JavaScript refers to a sequential set of statements that are combined into a compound statement.

v. Keywords such as *new, var, for* can also be used as variables' names.

2.7.3 Multiple-Choice Questions

i. When empty statements are encountered by the JavaScript interpreter:
   a. Error is thrown
   b. Simply ignores the statements
   c. Displays a warning
   d. Halts the execution of program

ii. Which of the following is not a data type belonging to JavaScript?
   a. Float
   b. Boolean
   c. Number
   d. String

iii. Which symbol is used for displaying multiline comments?
   a. \\
   b. * *\\
   c. * *\
   d. *\ *

iv. What is the correct output of the code snippet given next?

```
var grade = 'F';
var result;
switch (grade) {
    case 'A': result + = "10";
    case 'B': result + = " 9";
    case 'C': result + = " 8";
    default: result + = "Not applicable";
}
console.log(result);
```

a. 10

b. Not applicable

c. 098

d. None of the above

v. What will happen when the following code snippet is run?

```
var counter = 0;
while (counter <20) {
    console.log(count);
    count + +;
}
```

a. The implementation of *while* loop is incorrect.

b. The program stops execution abruptly.

c. The values of counter from 1 to 19 are calculated but not printed.

d. It prints the values 0,1,2...19 on the console.

2.8 Demo and Hands-On for Variables and Assignment Operator

2.8.1 Objective

i. Understand the purpose of a variable in JavaScript.

ii. Compare basic JavaScript data types.

iii. Write JavaScript code to declare and assign a value to a variable.

2.8.2 Prerequisite

- **Step 1:** Visit https://www.routledge.com/9780367641429 and download zip file for *Ch2_Variables_Template*.
- **Step 2:** Unzip the content and open *index.html* in any editor of your choice (example: Notepad or Visual Studio Code).
- **Step 3:** Insert given code next inside <script> block in *index.html*.
- **Step 4:** Repeat the above three steps for each code snippet given next.

2.8.3 Explore

Using any editor, we will write and test JavaScript code to create and test variables by viewing the results in the Google Chrome web browser. We will complete the following:

- Describe what a variable is and how variables are used in JavaScript.
- Write the code to declare a variable in JavaScript using correct syntax and naming conventions.
- Describe data types used in JavaScript and why they are important. Write code to compare variables in JavaScript using various operators.
- Practice assigning values to variables using the JavaScript assignment operator.
- Write a JavaScript code that uses a variable to collect data from a user and displays it on a web page.

2.8.3.1 Code Snippet-1

```
var p1 = "You are entering and storing all of this using a
variable"
var p2 = "A variable is a <u> Named location in memory </u> ";
var p3 = "Acts as a container that holds a data value";
var p4 = "The syntax to create variable is <u> var
variableName; </u> ";
```

2.8.3.2 Code Snippet-2

```
var p1 = "You are entering and storing all of this using a
variable"
var p2 = "A variable is a <u> Named location in memory </u> ";
var p3 = "Acts as a container that holds a data value";
var p4 = "The syntax to create variable is <u> var
variableName; </u> ";

var p1 = " You can declare (create) a variable using the var
statement. You just have to make sure to use correct syntax
for both the command and for the variable name. "
var p2 = " Observe the code and watch the comment for easy
understanding";
var p3 = " Try creating and manipulating variables";
var p4 = " use document.write (varName) to see the output in
Result section";
var x;                    //This is an undefined data type
```

```
var myNumber = 23;         //This is a number data type
var yourNumber = 45;
var myName = "Alex";       //Strings require double quotes
var myDog = "Google";
x = 5;                     //Assignment Operator
document.write(myNumber);         //used to print on page
document.write(myName);
document.write(myDog);
document.write("<br> ");          //used for new line
document.write(myNumber = = yourNumber); //Boolean
comparison
document.write("<br> The value of x is now " + x);
x = "blue";
document.write("<br> The value of x is now " + x);
var weeklyPay = 800;
document.write("<br> Weekly pay: " + weeklyPay);
weeklyPay + = 40;     //Assignment Operator with increment
document.write("<br> Weekly pay: " + weeklyPay);
```

2.8.3.3 Code Snippet-3

```
var p1 = "The numbers you keyed in were stored in variables
so that the JavaScript code could add them together and dis-
play the answer for you.";
var p2 = "In fact, the answer was also stored in a variable!
It took three variables to make this program work.";
var sum;
var a = prompt("Please enter a number:");
//take input from the user and store in variable
a = Number(a);
var b = prompt("Please enter a second number: ");
sum = a + b;
document.write("Thank you for entering " + a)
document.write(" and " + b + ". <br> ")
document.write("The total of your two numbers is: " + sum)
```

2.9 Demo and Hands-On for Control Flow Statements

2.9.1 Objective

i Understand control flow.

ii Use basic control flow operations.

iii Play with CSS and code logic using control flow.

2.9.2 Prerequisite

- **Step 1:** Visit https://www.routledge.com/9780367641429 and download zip file for *Ch2_ControlFlow_Template*.
- **Step 2:** Unzip the content and open *index.html* in any editor of your choice (example: Notepad or Visual Studio Code).
- **Step 3:** Insert given code next inside <script> block in *index.html*.
- **Step 4:** Repeat the above three steps for each code snippet given next.

2.9.3 Explore

Using any editor, we will write and test JavaScript code to interact with HTML/CSS code with logic and see the results displayed on the web page.

2.9.3.1 Code Snippet-1

```
// Hello there, I'm Snippet-1.

// Reference to content-box in HTML,
const myP1 = document.querySelector("#p1 p");
const myP2 = document.querySelector("#p2 p");

// Declare a few variables | Try playing with the va-
lues too.
let isJavaScriptCool = 1;

// A function to append content
const appendContent = (appendHere, msg) => {
    let myTextContent = document.createElement("p");
    myTextContent.innerText = msg;
    appendHere.appendChild(myTextContent);
};

/* Using control-flow, let's say you want to process some
piece of code conditionally.

   if, else | Syntax: if(<condition> ) {…True block…} else
{…Alternative/False block…} */

   if (isJavaScriptCool = == 1) {
    let msg = "True Block | | ";

       /* Try adding some code here, and see it being pro-
cessed, whenever condition is true. */
```

```
        msg + = " Appreciated man, Thanx!";

        // Append it to html, via above def. function:)

        appendContent(myP1, msg);
    }
    else {
        let msg = "False Block | | ";
        msg + = "May your workstation be Virus-infested.";

        // Append to HTML (#p1).
        appendContent(myP1, msg);
    }


    // You might ask, why '===' instead of '==' for comparison,
    // Using '==' javaScript will perform all sorts of coer-
cions (Type-Casting) to
    // equate the values, and will cause a bug which is very
difficult to identify, '===' is recommended.
    // Here's an instance,

    if (isJavaScriptCool = = "1") {
        /* Keep in mind, 'isjavaScript' is integer, and being
compared to string value "1" This shouldn't be true, but due
to '==' automatic coercion will take place */
        let msg = "True Block | | ";
        msg + =
            " Hello it's weird, I should not be printed, Don't
use '==' will cause bugs.";

        // Append to HTML (#p2)
        appendContent(myP2, msg);
    }
```

2.9.3.2 *Code Snippet-2*

```
    // Hello there, I'm Snippet-2.

    // Reference to content-box in HTML,
    const myInputBox = document.querySelector(".thoughts");
    const result = document.querySelector
(".results.content p:nth-child(2)");
    let resultColor = document.querySelector
(".results.content");
```

```javascript
    // Declare a few variables | Try playing with the va-
lues too.
    let isJavaScriptCool = 1;

    // A function to append content
    const appendContent = (appendHere, msg) => {
        let myTextContent = document.createElement("p");
        myTextContent.innerText = msg;
        appendHere.appendChild(myTextContent);
    };

    // Event-listener that triggers everytime something is
typed(Key-up) into target-inputbox.
    myInputBox.addEventListener("keyup", () => {
        // Length of string in text-input-box.
        const charCount = myInputBox.value.length - 1;
        result.innerText = charCount;

        // Lets use conditionals,
        // Assign different colors based on character-count.

        // Add more conditionals, for funny reactions.
        if (charCount <= 10) {
            resultColor.style.color = "green";
        } else if (charCount <= 20) {
            resultColor.style.color = "tomato";
        } else if (charCount> = 30) {
            resultColor.style.color = "red";
        }
    });
```

3

Objects

The quality of beauty lies on how beholder values an object.

—Toba Beta

In the initial phases of web development, the aim of a web page was to deliver static content to the users. A shift to dynamic programming brought more complexity in maintaining the code. The traditional way of maintaining variables and functions made things even worse. The simple variables holding values and individual functions were not helping, and a complete change of programming model was needed. Further, as more complex requirements for web pages arrived, it led to a change in perspective from procedural programming paradigm to the object-oriented programming paradigm.

Object-oriented programming (OOP) paradigm is adopted by most of the popular programming languages. The concepts of objects, classes, inheritance, polymorphism, abstraction etc. all combine to give the developer freedom to create more dynamic, well-designed, real-world solutions. Also, the produced code is easier to maintain, reuse and debug.

Objects are the fundamental building block in OOPs' way of writing code. To understand OOPs, we need a good understanding for the concept of object. Let us look at the physical world for the same. Everything around you is an object. Look at the table, chair, speaker, phone etc.; each of these objects can be represented by a *state* and a *behavior*.

A state defines a set of possible states any object can possess. At any point in time, it can possess only one of these possible states. While behavior defines how an object can migrate from one state to another, the behavior for an object can be understood as any action applied on the object. For instance, a table lamp can be in a state of *on* or *off* and have a behavior of getting turned on and off. Here, *on/off* are one of the possible states, and *turn on/off* are the actions performed to change its state. Similarly, a cat has her state defined in terms of color, weight, size and breed. A cat's behavior is judged by what she does, i.e. meows, runs, scratches etc. This is how the state and behavior for an object together make it look and act like any real-world object.

Quick exercise: Look at the physical objects around you and think about the different states they can be in; think about the different behavior they exhibit. If you were asked to

DOI: 10.1201/9781003122364-3

model those objects, then how would you describe these states
and behavior for those objects. Try for at least 10 different
objects.

This chapter discusses the concept of objects and how they differ from
other programming languages in terms of declaration and usage. A detailed
discussion on object's properties is provided followed by the concept of
classes. Further prototypal inheritance for JavaScript objects is discussed,
and the concept of inheritance with subclasses is discussed toward the end
of the chapter.

3.1 Objects

Objects are used to represent real-world entities, which makes programming a solution easier for the developer. Objects can store data in an organized and cohesive manner. Objects are the fundamental data type of
JavaScript [67]. Almost everything in JavaScript is object oriented, except
very few primitive data types; these primitives also have their wrapper
objects as we have already discussed in previous chapters. The objects are
variables that can contain multiple values as an unordered collection of
data. These can be changed or mutated by using the object's reference. An
object is used, denoted and handled by using the reference value assigned
to it. Reference values serve as the means to denote an object, which can in
turn contain multiple values.

Figure 3.1 represents the relationship among the program code, object
reference, object and its value. Program code uses the object reference to
access the actual object. The object can be considered as a storage box that
has the freedom to store any kind of value, either simple value or complex
aggregate values.

```
var obj = {}      //obj is the reference to an empty object
var obj2 = { "key1": "Value1", "key2": 1234}
//obj2 is reference to an object holding key1 and key2 with
their values
```

In the previous example, code *obj* is the reference name or handle name
used to denote the object. On the other hand, *obj2* can hold multiple values
such as string and number. The values inside *obj2* can be accessed as *obj2.key1*
(this will return *Value1* as string) and *obj2.key2* (this will return *1234* as
number). Objects allow the freedom to the developer to design and store data
in any way it is suitable. These named values are simply key-value pairs used
to store information and are generally called *properties* belonging to the object.

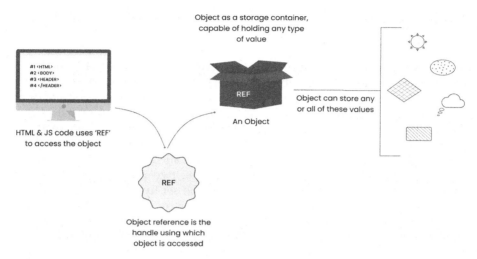

Object as a storage container, capable of holding any type of value

HTML & JS code uses 'REF' to access the object

REF

An Object

Object can store any or all of these values

REF

Object reference is the handle using which object is accessed

FIGURE 3.1
Object and its reference.

During development of a web page, a developer finds objects at various places such as a native object, a host object or a user-defined object. These are briefly explained as follows:

- *Native objects* are the ones defined by the ECMAScript specification such as arrays, functions and regular expressions.
- *Host objects* are the ones that are defined by host environment where the program runs, i.e. mostly the web browser. Browser Object Model and Document Object Model use host objects; these will be discussed in detail in upcoming chapters.
- *User-defined object*s are the custom objects created by the developer within the code as per the requirements. These are the most widely used objects that allow manipulation and execution of different operations, thereby governing the behavior of web page.

One of the special features of JavaScript is the concept of prototypes and prototypal inheritance. Every object in JavaScript inherits from the generic prototype called *Object*. This prototype can be easily referred by using *Object.prototype* invocation. All the built-in constructors for objects and most of the user-defined constructors inherit from object as their prototype. Objects can be further inherited by other objects, thereby making a connected series of objects, often called *prototypal chain*. Prototypal inheritance is further discussed in detail in upcoming sections. Let us now discuss the properties of objects in detail.

3.2 Properties of Objects

Objects have properties that are used to define as well as describe the state and behavior of a particular object. An object's property always has a name and an associated value with it. Name of a property is always a *string* or a Symbol (name, number or string literal) that is mapped to an actual value. The property for an object is similar to *HashSet* in Java language where a key should be unique, and hence, a value can be of any type, even a function. The property that has a function as a value is called a *method*.

An object must have distinct properties, i.e. two properties with same name cannot exist. Values of properties can be set by directly supplying a value as shown in next example for *person1* or by using *new* keyword as shown for *person2*. Also note that key will always be a string in a property. It does not matter whether you use " " or not; *first Name* is defined inside " ", while *age* and *gender* are written without " ". This is not an error but is a rather valid way to declare a key; we use " " only when the key is *multiword*. Similarly, in case of an object *person2* creation using the new keyword, we can add any type of property and any number of properties. The property key defined using *person2["first Name"]* refers to the associated key here.

```
var person1 = { "first Name" : "Ram" , age : 50 , gender : "M" }
//static way to create object
var person2 = new Object () ;
//dynamic way to create object
person2 ["first Name"] = "Ram" ;
// for multiword key use square bracket []
person2 . age = 50 ;           // can add properties later
person2 . gender = "M"      // any type of properties
```

Dot notation is used to access and represent the objects and their properties such as *objectName.propertyName*. In case a key is multiword, square brackets are used. Objects are mutable, i.e. changeable in nature and are manipulated using the reference or handle name. Each property contains some additional metadata; let us understand in the following sub section.

3.2.1 Property Configuration Descriptors

We have seen that every property contains a key and value. This is a very broad perspective; to be more specific, the key is called *name* of the property, and the value part is one of the attributes among four others. The four attributes for any property are *value, writable, enumerable* and *configurable*. These attributes govern how a given property behaves when it is accessed, read or written to. All the attributes of a property are readable in JavaScript, by default. These attributes are further explained as follows:

- *Value attribute:* The value attribute is the *data* part that a property will store and is used as it is. The rest of the attributes define how a value or a property can be accessed.

- *Writable attribute:* It is used to check whether the value of an object property can be *assigned* to a new value or not. This is by default *false* for most of the in-built objects and *true* by default for user-created object.

- *Enumerable attribute:* It is used to check whether the property name is used or returned by loops (while/for), i.e. whether iteration or enumeration is allowed or not. When set to *true*, property is allowed to enumerate using *for...in* loops (discussed later). This is also by default *false* for most of the in-built objects and *true* by default for user-created object.

- *Configurable attribute:* It specifies whether the configuration of a given object can be *altered* or not. By default, this attribute is set to *true*. If it is set to *false*, a property's configuration cannot be modified.

The next example demonstrates a newly created custom object *book* and the static method of Object. *Object.getOwnPropertyDescriptor* is used to see the property config descriptors. Observe the output shown in the comments section.

```
var book = {
    title: 'Harry Potter',
    author: 'J.K. Rowling',
};
console.log(Object.getOwnPropertyDescriptor(book,
'title'));

/* Output will be the following object:

value: "Harry Potter"
configurable: true
enumerable: true
writable: true
*/
```

Table 3.1 displays the attribute, type and default values for the in-built and user-defined types.

Most of the built-in object properties are nonwritable, nonenumerable and nonconfigurable. Properties defined on user-created objects are mostly writable, enumerable and configurable.

TABLE 3.1

Attributes of a Property

Attribute	Type	Default Value for In-Built Objects	Default Value for User-Created Object
Key	String or symbol type	Undefined	User-defined key
Value	Any data type	Undefined	User-defined value
Writable	Boolean	False	True
Enumerable	Boolean	False	True
Configurable	Boolean	False	True

Relationship between object properties is hierarchical in nature. JavaScript does not restrict developers to work with a single level of properties for objects. Instead, as mentioned earlier, everything is an object in JavaScript. Therefore, it allows the properties of objects to have their own properties, further down the hierarchy. This way the data structures can be dynamically expanded while programming. This also means that objects can be nested within each other to provide a flexible programming style. This is a very powerful feature allowing creation of objects within objects with extended properties and methods.

3.3 Creating Objects

JavaScript allows creation of objects in isolation, i.e. without making classes. Primarily, there are three ways to create an object in JavaScript. An object can be created by using literals, the *new* keyword or by using *Object.create()* function [67].

- *By using literals:* An object literal is a comma separated list of colon (:) separated key-value pairs where each pair signifies an object property. Object literal creates and initializes a new object, each time it is encountered. Every time an object literal is parsed or evaluated, it creates an object with some predefined configuration properties and supplied value.

```
let emptyObject = {};    //empty object with no properties
let line = { x: 6, y: 10 };
/*line object with properties of x and y */
let room = {
    walls: 4, flooring: tiles,
    owner: { first_name: Jane, last_name: Doe }
```

```
    };
/* Declaring a nested object room containing owner object
inside it*/
```

In the code snippet, we can observe that a key for an object can be a string or a Symbol or a number, and the corresponding stored value can be of any type. Object *room* has properties *walls*, *flooring* and *owner*, where *owner* is an object itself with properties *first_name* and *last_name*. An object can have another object as its property, and this is called nesting of objects. This allows creation of a hierarchy of objects, which empowers the developer to expand the objects. Placing object literals within recursive statements will thus create a new object each time the object literal is evaluated.

```
var obj = new Object();
//create object similar to var obj = {};
var arr = new Array();
//create object similar var arr = []
var dt = new Date();
//creates date object
```

- *By using new keyword:new* operator creates and initializes an object. Invocation of *new* keyword must be followed by a method invocation. This special method is called *constructor*, primarily used for creation and initialization of new objects. JavaScript has its own built-in constructors for its built-in data types. Apart from built-in constructors, custom constructors can be defined to initialize objects.

obj, arr and *dt* are empty objects created using the *new* keyword; these objects inherit from generic Object, Array and Date prototypes, respectively. Apart from built-in constructors, custom constructors can also be defined to initialize the objects.

In the following code snippet, a new object is created using the *new* keyword, and it is initialized using a custom constructor defined using *function*. This type of constructor is called as object constructor function, and the Person object will be derived from Object class automatically.

```
function Person(fName, age, gender) {

    this.fName = fName;
    this.age = age;
    this.gender = gender;
    }
    var myMother = new Person("Alice", 51,"F");
                            //myMother object is created
```

- *By using Object.create() function:* This method can be used to create a new object. It can also take an optional argument, i.e. to create an object based upon the prototype object. *Object.create()* method is a static method of the Object class. *Object.prototype* will create an object that inherits basic methods from the prototype. Passing arguments within the *Object.create()* are considered as properties and must be supplied in key-value pairs. These properties are inherited by the newly created object. Passing *null* as an argument will create an empty object that will not hire anything from *Object.prototype.* Let us understand this using the example given next.

```
var obj1 = Object.create();
//creates an empty object
var obj2 = Object.create(null);
//null object does not inherit anything from prototype
var obj3 = Object.create({ x: 1, y: 20 });

/*Object created with name obj3 and inherited properties
x and y */

var obj4 = Object.create(Object.prototype);
//creates an ordinary object equivalent to obj1
var obj5 = Object.create(myMother);
//creates a copy of myMother object created in previous
example

var obj6 = Object.create(Person.prototype);
//creates empty object
```

As we can understand from the example, *Object.create()* method is a static method, and it can be used in a few different ways. *obj1* is an example of creating an empty object, and the object created is similar to creation of an object using *var obj1 = {}.* The other objects *obj2, obj3, obj4, obj5,* and *obj6* are created by passing an argument inside *Object.create()* method. The arguments passed ensure the newly created object comprises the properties of the passed *object, properties* or *prototypes.* Passing *null* in case of *obj2* is an exception to this statement, as *null* is a primitive data type that does not have any property.

3.4 Objects as Record and Dictionary

Objects are created either with a fixed number of properties or with a variable number of properties. Objects with fixed number of properties

TABLE 3.2

Difference between Objects: As Records and as Dictionary

Objects as Records	Objects as Dictionary
Fixed number of properties that are specified beforehand	Dynamically changing or variable number of properties that can be expanded as per need
Keys are known at the time of development	Keys are unknown at the time of development
Values can have different data types	All values have the same data type

declared at development time are called object as records, and the objects with a variable number of properties, which can be dynamically changed, are called object as dictionaries [68].

Objects work best as records. But before ES6, JavaScript did not have a data structure for dictionaries (ES6 brought Maps) [69]. Therefore, objects had to be used as dictionaries, which imposed a significant constraint: keys had to be strings (symbols were also introduced with ES6) [70] (Table 3.2).

When creating objects as records, one must specify the key-value pairs of information that the object is going to use. It allows storing properties with different data types and is, therefore, used to create objects that need to store specific but with varied properties. On the other hand, using objects as dictionary allows the code to dynamically add as many properties as needed, but all of them must have the same data type. There is no need to specify every property beforehand; rather properties can be added as and when needed throughout the execution of program.

3.5 Operations on Objects

Properties of an object can be accessed, mutated or iterated by using certain predefined functions. In this section, we will discuss how to *set* and *get* an object's property. These are called as setter and getter methods, more commonly known as *accessor* and *mutator* methods. We then elaborate on different functions that can be used to check on various object properties and property configuration descriptors.

3.5.1 Accessors and Mutators

To access or mutate any object property, a dot operator or square brackets are used. A property can be accessed by simply using the object name, followed by the dot operator with the property name, or passing the key inside square brackets. The latter approach is used for the multiword key.

Similarly, dot and square brackets can also be used to set an object's property. The following example explains.

```
var book = { }          //creating an empty object of book

    book.title = "Harry Potter"
    //using. operator to set value of title

    book["authored by"]="J.K.Rowling"
    //using square bracket to set a multiword key property

    var title = book.title;
    //returns the title property of book object using. operator

    var name = book["authored by"]
         //returns the authored by property of author object
using []
```

The approach discussed in the previous example is very straightforward. JavaScript provides us a way to define getter and setter accessors that are invoked by getting/setting a property.

A getter is a method-like entity that is invoked by getting a property; observe in the following example that *full* is a method created by adding a *get* prefix. The call to *book.full* will return Harry Potter-J.K.Rowling. Observe that we are not using () after invoking *full* because this is not a method call but a special get property.

```
    var book = {
        title: "Harry Potter",
        author: "J.K.Rowling",
        get full() {
            return `${this.first}-${this.last}`;
        },
    }; //Created a book object with get method.
```

Similarly, a setter is a method-like entity that is invoked by setting a property; observe in the following example that a method *full* is prefixed with *set* and *details* are passed in. The method splits the details on '-' and sets title and author property. Further when *book.full* property is set, the *full()* method is invoked with *details* and arguments.

```
    var book = {
        title: "no title defined",
        author: "no author defined",
        set full(details) {
```

```
        var parts = fullName.split('-');
        this.title = parts[0];
        this.author = parts[1];
    },
}; //Created a book object with a setter/accessor method.
```

```
book.full = "Harry Potter - J.K.Rowling" //setting the
property of full
console.log(book.title) // Harry Potter
console.log(book.author) //J.K.Rowling
```

3.5.2 Useful Operations and Loops

JavaScript allows the functionality to test whether the properties exist inside an object or not. This can be achieved by any of the following methods:

- *Assignment operator:* The assignment operator is used to check the existence of property. If the property is not present inside the object, *undefined* is returned. The next example demonstrates the usage of the assignment operator.

```
var book = { title:"Harry Potter", author:"J.K.Rowling" };
//created a book object
```

```
console.log(book.title!== undefined)      //prints true
console.log(book.author!== undefined)     //prints true
console.log(book.isbn!== undefined)       //prints false
```

- *Using the in operator:* The *in* operator is a binary operator that checks the left operand property inside the right operand object; this returns true if the object possesses that property defined itself or in its inherited properties. In the next example, we can see that we get *true* for *toString* property also; this is because of inheritance.

```
var book = { title:"Harry Potter", author:"J.K.Rowling" };
//created a book object
console.log("title" in book)          //prints true
console.log("author" in book)         //prints true
console.log("isbn" in book)           //prints false
console.log("toString" in book)
//prints true as toString is inherited property.
```

- *hasOwnProperty():* This special method is provided to check for inherited property and own property defined on the object. This method returns *true* for the object's own properties and returns *false* for the inherited properties. In the example given next, we can see that we get *false* for *toString* property (this is because *toString* is an inherited property).

```
var book = { title: "Harry Potter", author: "J.K.Rowling" };
//created a book object

console.log(book.hasOwnProperty("title"))
//prints true
console.log(book.hasOwnProperty("author"))
//prints true
console.log(book.hasOwnProperty("toString"))
//prints false
```

- *propertyIsEnumerable():* The method *propertyIsEnumerable* method is more refined version of *hasOwnProperty* method. It returns *true* only if the property is object's own property and the enumerable attribute is set to *true*. It returns *false* for the properties whose enumerable attribute is set to *false,* i.e. iteration over those properties is not allowed. Next is the example to demonstrate its usage.

```
var book = { title: "Harry Potter", author: "J.K.Rowling" };
//created a book object

console.log(book.propertyIsEnumerable("title"))
//prints true
console.log(book.propertyIsEnumerable("author"))
//prints true
console.log(book.propertyIsEnumerable("toString"))
//prints false
```

- *Iterating over properties:* Now, it's a good time to understand the use of special *for...in* loop, which we have discussed back in Chapter 2. Now you understand that an object contains many properties, and there might be the requirement to iterate over the keys of that object. In all those cases, the *for...in* is used. The syntax has been discussed before, so let us understand this with an example for the *for...in* loop to set it in action.

```
var book = { title: "Harry Potter", author: "J.K.Rowling" };
//created a book object

delete book.title;
//Deleting the title property from the book object
console.log(book)
```

```
delete book; //No effect on object
console.log(book)

/*

Output will be:
{author: "J.K.Rowling"}
{author: "J.K.Rowling"}
*/
```

Always remember that *for...in* will iterate over all the properties, including the ones that are inherited and are enumerable. Generally, all the built-in properties or the inherited properties of an object are not enumerable in nature, while the user-defined properties are added and enumerable by default. In such cases, *hasOwnProperty* method can be used to filter out inherited properties inside the *for...in* loop.

There are many methods available that provide us keys and values based upon the requirement that can be used to iterate over.

- *Object.keys()* method returns an array holding the object's enumerable properties.
- *Object.getOwnPropertyNames()* returns an array of all properties of string type belonging to the object, irrespective of its enumerable status.
- *Object.getOwnPropertySymbols* returns an array of all properties of symbol type belonging to the object, irrespective of its enumerable status.
- *Reflect.ownKeys()* returns all the properties belonging to an object including enumerable, nonenumerable, strings and Symbols.

- **Deleting object properties:** At the time of development, one may need to delete the property of an object. The *delete* keyword can be prefixed before an *object.propertyName* or *object[propertyName]*. It then deletes both the property name (key) and the attribute (value). Once deleted, the property is not available to be used until you add it again. The delete operator has no impact on variables and function; it is only used for property of an object. Also, exercise caution while using delete with the predefined objects and their properties. As deleting their property may leave your application in a suspended or, worse yet, crash state.

```
var book = { title: "Harry Potter", author: "J.K.Rowling" };
//created a book object
```

```
for (key in book) {
    console.log("Key is: " + key + " Value is: " + book[key])
}

/*
Output will be:
Key is: title Value is: Harry Potter
Key is: author Value is: J.K.Rowling
*/
```

On successful deletion of a property, true is returned else *false* is returned. If the property being deleted does not exist, then it simply returns *true*. If a property with the same name is present within the object's prototype chain, then the user-defined property is deleted and the property existing within the prototype chain is used by the object.

If the property attribute is set as a nonconfigurable property, then delete will simply return *false* and a *TypeError* is thrown. Nonconfigurable properties belonging to built-in objects such as Array, Math and Object cannot be deleted using the delete keyword [71].

Delete keyword simply breaks the reference of an object from the property. It does not free the memory as observed in other programming languages. Once a property is untied from an object reference, it is automatically deleted. Memory management in JavaScript is executed via breaking references from the associated properties. Such hung properties unassociated with any object are deleted, and the memory is freed.

3.5.3 Object Methods

We have already discussed about the state of an object. In this section, we will discuss the behavior of an object. Object methods define the actions that can be performed on objects or objects' properties. An object method is simply an object's property containing a method declaration and definition. They can be considered as a key-value pair where key denotes the method name being defined on the object and value denotes the method declaration.

```
let person = new Object();       //creating a new object
person.fName = "Thomas";         //adding property 'fName'
person.lName = "Shelby";         //adding property 'lName'
person.fullName = function ()
//property assigned a function
{
    console.log(this.fName + this.lName);
    //usage of this keyword
```

```
}
//we call such property as method
```

In the previous example, we created an object person and defined three properties: *fName, lName* and *fullName.* The first two properties should be clear to you by now; the third property *fullName* is assigned to a *function(),* and this assignment is called *method.* The body of method can access all other properties of the object using a special reference variable *this. this* keyword is used to represent the owner object, i.e. in this example, *this* refers to *person* object. Another way to create the same *person* object and add a method to an object is shown next:

```
let person =                //creating object
{
    fName: "Thomas",        //defining properties
    lName:"Shelby",
    fullName: function () { //property assigned a function
        console.log(this.fName + this.lName);
        //usage of this keyword
    }                       //we call such property as method
}
```

In both the examples, the *person* object is created with its own properties and method that can be accessed simply by using *objectName.methodName(),* for instance, *person.fullName().* The parenthesis '()' are used to depict the method calling/declaration.

3.5.4 Built-In Object Methods

All JavaScript objects that are created from a prototype inherit properties as well as methods from *Object.prototype.* Inherited properties are methods defined on *Object.prototype.* These methods are meant to be replaced by more specialized implementations by the newly created object. JavaScript allows easy copying of properties of one object to another object such as:

```
let obj1 = { a: 1, b: 6 }
let obj2 = { c: 8, d: 10 }
for (let key of Object.keys(obj2))
//iterating over every key of obj2
{
    obj1[key] = obj2[key]; //copying to obj1 from obj2
}
//Output:
//obj1 is: {a: 1, b: 6, c: 8, d: 10}
```

The same operation can be done using the *extend()* utility function provided by JavaScript. Another way is to use *Object.assign()* function with two or more objects as passed arguments. First argument is called the *target object*, which is modified and returned. The second argument is the *source object*, which is left unaltered. *Object.assign()* uses the simple setter and getter methods by invoking them to retrieve and set the property values in source and target objects, respectively.

```
Syntax:
Object.assign(target object, source object);
//overwrites every property of source object to target object
```

example:

```
let obj1 = { a: 1, b: 6 };
var obj2 = { c: 8, d: 10 };
Object.assign(obj3, obj1)
```

- *toString()* method returns the string value of the object property without taking any arguments. It is primarily used by JavaScript while using operators with objects such as + , - operators. It can be used where a string is expected instead of an object. By default, very limited information is passed by the *toString()* method. Therefore, this method is defined by many classes on their own.

```
let line = {
    x: 2,
    y: 5,
    toString: function () {
        return '(${ this.x }, ${ this.y })';
    }
};
```

- *toLocaleString()* method returns the localized string representation of the calling object. By default, this method simply calls the *toString()* and returns the value. Classes can define their own implementation for this method.
- *valueOf()* method is used to convert an object to another primitive data type instead of a string. Many built-in classes provide their own implementation for the method. Whenever primitive data values are required from an object instead of String, this method can be implemented by the class.

3.6 Prototypal Inheritance

Programming languages like PHP, Java and Python are class-based languages and use classes for object creation. JavaScript is a class-based as well as a prototype-based language, which offers prototypes and later supported classes to create objects. Using prototypes, different properties can be shared through generalized objects that can be further cloned, extended or inherited [72]. Extending functionality using prototypes is called as prototypal inheritance. Every object in JavaScript inherits from prototypes either directly or indirectly.

Apart from having its own set of properties, every object in JavaScript inherits its properties from another object. Generic objects from which the properties and methods are inherited are referred to as the *prototype*. For instance, all objects inherit from the same prototype object, i.e. *Object.prototype*. All the created objects in JavaScript have *Object* as its prototype. It is the final parent for all objects, and accessing anything beyond it will result in *null*.

As depicted in Figure 3.2, every other prototype inherits from *Object.prototype*. Any type of array, function, regular expression, Number, String or user-created object inherits from *Object.prototype*. This allows calling the properties and methods defined in Object prototype by all the other objects. Using prototypes thus helps in adding more functionality to objects without the need of declaring them with every class or object.

Every object also has a hidden internal property called *__proto__*, which is inaccessible directly in code. This hidden property can be accessed by using the provided static method of Object prototype, i.e. *Object.getPrototypeOf (obj)*. This method returns the list of all the built-in properties and methods belonging to the passed object.

```
var obj1 = {};         //empty object, inherits nothing
var obj2 = new Object()    //empty object, inherits from
Object prototype
Object.getPrototypeOf(obj2);
```

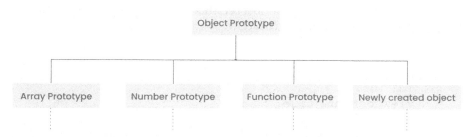

FIGURE 3.2
Hierarchy of inheritance via *Object.prototype*.

```
//returns a list of properties and methods inherited from
prototype
```

Properties defined directly for an object are referred to as *own property* or *noninherited property*. Properties inherited from prototypes are referred to as the *prototypal properties.*

3.6.1 Prototype Chaining

Every object contains the [[Prototype]], i.e. __proto__ object, which points either to another object or to null. Imagine an object Z with .__proto__ property set to ObjectY and further ObjectY with .__proto__ property set to ObjectX. This way prototypes inherit from one another as and when needed. One prototype can be linked to another that can further be linked to another, building a prototype chain. As mentioned before, newly created objects inherit from one prototype or another. These prototypes inherit from one another leading to creation of chains of prototype whose root is always *Object.prototype.* The following example explains this:

```
var X = { key1: "value1" }
var Y = { key2: "value2" }
Y.__proto__ = X
var Z = { key3: "value3" }
Z.__proto__ = Y
```

The screenshot in Figure 3.3 demonstrates the execution of above code. You can observe the output expanded; the Z object now contains Y inside its *__proto__* object, and Y object contains X object inside its *__proto__* object. This ensures an inheritance of sharing properties, i.e. the object Z has access to *key1, key2 and key3.*

Similarly, the prototypal inheritance can also be used to inherit methods from other prototypes; for example, in the code snippet given next, the *youTubePremiumUser* is inheriting properties youTubeUser through *__proto__* object and methods. You are encouraged to try executing this code on your machine, and you will observe that *youTubePremiumUser* inherits the getEmail method from *youTubeUser.*

```
let youTubeUser = {
    email: "dummyemail@gmail.com",
    id: 101,
    viewContent: true,
    set setEmail(email) {
        this.email = email;
    },
    get getEmail() {
```

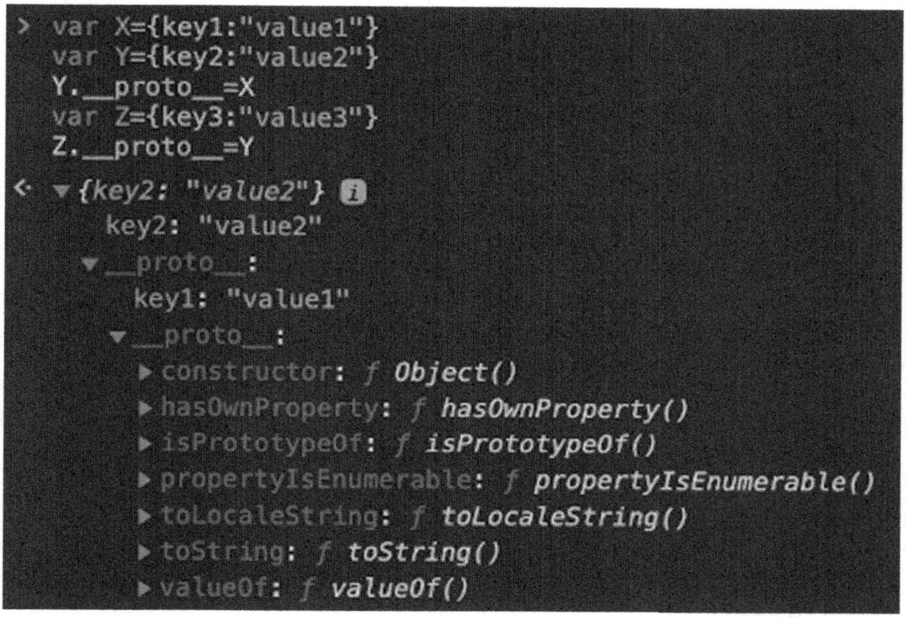

```
> var X={key1:"value1"}
  var Y={key2:"value2"}
  Y.__proto__=X
  var Z={key3:"value3"}
  Z.__proto__=Y
< ▼ {key2: "value2"} ⓘ
    key2: "value2"
    ▼ __proto__:
      key1: "value1"
      ▼ __proto__:
        ▶ constructor: ƒ Object()
        ▶ hasOwnProperty: ƒ hasOwnProperty()
        ▶ isPrototypeOf: ƒ isPrototypeOf()
        ▶ propertyIsEnumerable: ƒ propertyIsEnumerable()
        ▶ toLocaleString: ƒ toLocaleString()
        ▶ toString: ƒ toString()
        ▶ valueOf: ƒ valueOf()
```

FIGURE 3.3
Output screenshot.

```
            return this.email;
        }
    };
    let youTubePremiumUser = {
        ads: false
    };
    youTubePremiumUser.__proto__ = youTubeUser;
    //youTubeUser is a defined prototype
    console.log(youTubePremiumUser.viewContent);
    //returns true
    console.log(youTubePremiumUser.getEmail);
```

3.7 Classes

You must be wondering why we are discussing classes after objects. Well, JavaScript classes were introduced in the language specification ES6 in 2015 [73]. Before 2015, every object had an internal property called *[[Prototype]]*. This prototype was used to provide different properties and methods to the newly created object. With the introduction of classes, a clean and more elegant way to empower objects emerged. Classes allow the objects to operate in a safe and clean environment without getting lost in different prototypes and their methods.

Object prototype is the generic prototype, and every object inherits from it, regardless of it belonging to a class or not. Therefore, it is not mandatory for objects to always belong to a class in JavaScript. An object can exist independent of a class. A class declaration and definition thus become a matter of choice from a developer's perspective.

Coming back to the question, *what is a class?* In simple terms, a class is a blueprint or a template to generate objects. A class is used as a template to combine objects, its properties and the methods to operate on them. Addition of classes to JavaScript allows developers to easily operate and move between other OOP languages [74].

Class in JavaScript can be defined by using *class* declaration. The keyword *class* is used before the name of the class, and it is followed by the declaration statements. The syntax for creating a class is given next.

```
Syntax:
class MyClass {
  // class methods
  constructor() {… }
//Initialize properties

  method1() {… }
//define methods
  method2() {… }
  method3() {… }…
  …
}
```

Classes in JavaScript are strictly checked for syntax to increase system performance. Classes operate in a *strict mode* with strong syntax checking, and this cannot be changed, i.e. classes will always work in strict mode. A class must be declared first and then used; vice versa generates a *ReferenceError*. Class methods are *nonenumerable* in nature. Although classes are considered as a special type of functions, such restrictions are not present in functions and are only applicable for classes. As a naming convention, the first letter of class name is always capitalized, which helps in distinguishing a class name within the program.

3.7.1 Constructor, Properties and Methods

A class can contain different properties and methods depending upon its design and use. A class is declared using the *class* keyword and can be

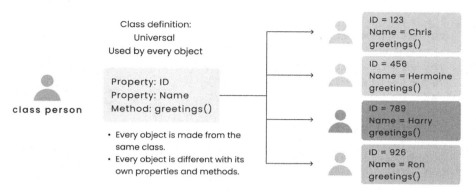

FIGURE 3.4
Class and object.

declared only once. Class definition consists of constructor and methods. The properties are initialized using constructors, and these properties along with the methods automatically become a part of the object initialized using this class. Every object will have its own copy of these properties. Every object created using this class will be different but still same in nature, i.e. properties (state) and methods (behavior) will be same for every object.

A class will always have its own constructor. Constructor is used to initialize the values of the newly created objects. It is a special method for creating as well as initializing an object, which is instantiated using a class. There can be only one constructor for one class. Every time a new object is created, constructor in the class in invoked to initialize properties for that object. If more than one constructor definition is present in a class, a *SyntaxError* is thrown.

In Figure 3.4, a class *Person* is shown and is used to create four objects. Every object is distinct yet derived from the same base class. The class declaration can be done as demonstrated next:

```
class Person                  //declaring a class
{                             //defining the class
    constructor(uid, name)    //constructor for the class
    {
        this.uid = uid;       //usage of this keyword
        this.name = name;
    }
    greetings()               //adding a method to class
    {
        console.log(`Hello ${this.name}`);
    }
}
```

```
}
var e1 = new Person(123,"Chris");
//creating e1 object from class Person
var e2 = new Person(456,"Hermoine");
var e3 = new Person(789,"Harry");
var e4 = new Person(926,"Ron");

e1.greetings();              //calling method on e1
    /*
Output will be:
Hello Arthur Doyle
        */
```

In this example, a new class is created with a single constructor and two properties, i.e. uid and name. A method is also provided *greetings()* along with a constructor to initialize properties. When the method is called as a property by the object, it executes and displays the output.

3.7.2 Extending Classes

Classes are helpful in designing templates beforehand that can be used later on. However, at times, a developer wants to modify or update the class definition to add some properties or methods. This can be easily done by extending the classes to add new properties. In the previous example, a person object is created with *uid, name* as properties and *greetings()* as a method. If a person object is needed with additional properties such as *lastName, age, gender* and methods *getAge(), getGender(), getfullName()*, then how can we use the same class to add these properties? Answer to this lies in extension of classes.

Classes can be extended to add additional functionality to existing classes. Any class can be extended to add more properties and methods to enhance its use. A class can extend a parent class using the keyword *extends*. A child class can call the constructor of the super class by using the keyword *super*, which calls the parent class constructor. In the following example, the previously used *Person* class is used as a base class to create a new *Student* class with more properties and methods.

```
class Student extends Person      //extending from parent
{
    constructor(uid, name, age, gender, lastName)
    //constructor
    {
        super(uid, name);
    //chaining constructor to call parent
        this.age = age;
```

```
        this.gender = gender;
        this.lastName = lastName;
    }
    getAge() //adding more methods to class
    { return this.age; }
    getGender() { return this.gender; }
    getfullName() { return (this.name + this.lastName); }
}
var e5 = new Person(123,'Lucy');
e5.greetings();          //displays 'Hello Lucy"
var s1 = new Student(3674,"Anne", 24, F,'Hathaway');
stuObj.getAge();          //returns 24
stuObj.greetings();       //displays "Hello Anne"
stu.Obj.getfullName();    //returns "Anne Hathaway"
```

3.7.3 Getters and Setters in Classes

We have already discussed accessor and mutator methods applied on objects; the very same concept is applicable for classes also. Just prefixing the *get* or *set* before the property name enables the accessor/mutator methods, i.e. when the property is called, the corresponding getter/setter method is invoked that returns/sets the value of the property.

Using getter and setter methods instead of directly operating on object properties is a safe and secure alternative. It preserves the integrity of data being retrieved and manipulated. These methods can be defined by using the keywords *set* and *get* as discussed before. The example given next explains their usage in action.

```
class Person {
    constructor(name, age) {
      this.name = name;
      this.age = age;
    }
    get age()        //Notice space between get and age.
    {
        return this.age;
    }
    set age(x) {
        this.age = x;
    }
}
//creating a new object and accessing properties using get
method
  var obj = new Human('Tony', 42);
  console.log(obj.age);
```

```
//calling get age() through property name
obj.age = 56;
//calling set age(56) through property name

console.log(obj.age);              //displays '56'
```

3.7.4 Static Members

Sometimes, it is not necessary to create a method or a variable for each object, and to save memory, we only need one copy of that property. In such cases, JavaScript offers the use of static members. The static member belongs to the class and not to individual objects. Therefore, it stays with the class, and instantiated objects can only access this property, but they never own this property [75].

Both property and methods can be declared static, and the static members are inherited by the child class. Static members can be added by using the keyword *static* as a prefix for property or method during class declaration. The static member can also be added later on using the *ClassName.property = 'value'* to an already-defined and declared class. The next example demonstrates the usage of static property and method.

```
class Account {
   static rateOfIntrest = 10 //static property
   static totalAccountHolders = 0 //static property
   constructor(accountNumber, branch, balance) {
       this.accountNumber = accountNumber
       this.branch = branch
       this.balance = balance
       Account.totalAccountHolders + +
       //incrementing static member when new object is created
   }
   static getNumberofAccounts() //static method
   {
       return Account.totalAccountHolders
   }
}
var client1 = new Account(12345, "Central Office", 42000)
var client2 = new Account(67891, "West Office", 12000)
console.log(Account.rateOfIntrest)
//call to static property
console.log("Total Accounts" + Account.
getNumberofAccounts())
//call to static method
   /*
```

```
Output will be:
10
Total Accounts 2
*/
```

Static methods and properties are inherited by objects that extend these classes. The Inheritance concepts remians same for normal and static members of the class.

3.8 Garbage Collection

We have learned *delete* operator on objects before; let us revisit again with some other questions like, what garbage collection is and why do we need it? Garbage collection is the process of freeing the allocated memory when it is no longer needed by the program. JavaScript offers automatic garbage collection that periodically cleans the memory and de-allocates the memory region [71].

To manually remove or delete a property from an object, *delete* operator can be used. Delete operator does not simply operate on the value of property but rather deletes the actual property value. For example:

```
delete book.author;    //delete property author from book
delete book["isbn"];   //delete property isbn from book
```

Delete operator is used to delete the object's own properties. It is not recommended to delete the inherited properties. To delete the inherited properties, they must be deleted from the prototype where they are defined. Deleting properties from prototypes affects every object that inherits from that particular prototype. Also, we should keep in mind that the use of *delete* does not remove the properties whose configurable attribute is set to *false*. Delete operator returns *true* when delete operation succeeds.

So, how to delete the complete object and let garbage collector do its job of freeing up the allocated memory? The only way to free up the memory is to mark the object as ready to delete; this can be done by setting the reference of the object to *null*.

Many a times, object references are passed, assigned and reassigned multiple times. This creates a complex relationship between actual objects and their references. Once a reference is unreachable, it will be deleted. One must be careful before removing or reassigning these references.

Garbage collection in JavaScript uses the algorithm *mark and sweep* for de-allocating memory. Garbage collector traverses the current execution

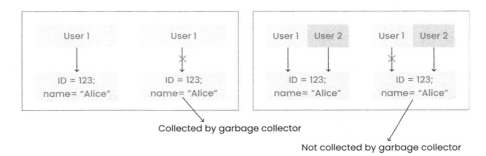

Collected by garbage collector

Not collected by garbage collector

FIGURE 3.5
Automatic garbage collection.

function (with its local variables and methods), other nested function, global variables etc. Every unreachable object is *marked*. Marked objects and their associated references are traversed and *marked* as needed (visited objects are remembered by garbage collector to avoid parsing same objects again and again). These references are traversed until every reachable and unreachable value is traversed. Once done, all the *marked* objects are removed, and the process initiates again. Garbage collection process occurs on its own in an automated manner. It cannot be forced to execute or stopped from execution. All the reachable objects are retained in memory. To understand the scenario, let's look at the next example and corresponding Figure 3.5.

```
class Human {
var name;
var ID;
Constructor(id, name)
{
    this.name = name;
    this.ID = ID;
}
}
var user1 = new Human(123, "Alice");  //creating new object
user1 = null;    //resets the reference to null & marks for
deletion
console.log(user1.name);    //does not work
```

In this code snippet, the object user1 is created using the constructor and is set to null. Setting the reference value to *null* makes it ready for garbage collector to delete. Further, a second scenario demonstrates that if a single reference is a variable referring to the object, then it will not be deleted or marked to be deleted. This second scenario is demonstrated as follows:

```
var user1 = new Human(123,"Alice");  //creating new object
user2 = user1;           //storing reference to user2
user1 = null;            //resetting reference to null
console.log(user2.name);  //it works
```

3.9 Exercise

3.9.1 Theory

    i. Differentiate between classical inheritance and prototypal inheritance.

    ii. How many programming paradigms are used by JavaScript? Explain.

    iii. What do you understand by prototypal inheritance in JavaScript? Discuss the use case when a developer should use prototypal inheritance.

    iv. Enumerate different ways of creating objects.

    v. Discuss different ways of iterating over objects. Critically analyze every possible alternative, and brainstorm which one should be used when.

    vi. Write a program to iterate over the object's properties, and display them on the console.

3.9.2 True/False

    i. A prototype chain is chain created by linking a set of objects together.

    ii. The method isPrototypeOf() is used to determine whether an object is a prototype of another object.

    iii. Global variables take precedence over the values of local variables.

    iv. *toString()* method is used on objects to get the numerical representation of the objects value.

    v. The ternary operator is invoked using the '?' in JavaScript.

3.9.3 Multiple-Choice Questions

   i. During the different stages of developing a website, a web developer encounters which of the following objects?

     a. User-defined objects

     b. Host objects

     c. Native objects

     d. All of the above

  ii. An object can be invoked using which of the following keywords?

     a. This

     b. Of

     c. Object

     d. None of the above

 iii. When a class 'C' extends another class 'D' in a program, we can say that class 'C' is the _____ class and class 'D' is the _____ class.

     a. super, sub

     b. parent, child

     c. Both a and b

     d. None of the above

 iv. What is the output of the following program code?

```
function newPerson() {
    this.name = 'Ojo';
}
function obj() {
    obj.call(this);
}
obj.prototype = Object.create(newPerson.prototype);
const newObject = new obj();
console.log(newObject.name);
```

     a. Throws error

     b. Does not compile

     c. Prints 'Ojo'

     d. None of the above

v. What is the output of the following program code?

```
const constantOne = { };
const constantTwo = Object.create(constantOne);
console.log(Object.getPrototypeOf(constantTwo) ===
constantOne);
```

    a. False

    b. True

    c. Compile time error

    d. 0

3.10 Demo and Hands-On for Objects

3.10.1 Objective

    i. Create a JavaScript Object and learn about properties.

    ii. Create objects with different methods and using getter/setter methods.

    iii. Create objects with user-defined Object Constructor and some action on it.

3.10.2 Prerequisite

- **Step 1**: Visit https://www.routledge.com/9780367641429 and download zip file for *Ch3_Object_Template*.
- **Step 2**: Unzip the content and open *index.html* in any editor of your choice (example: Notepad or Visual Studio Code).
- **Step 3**: Insert code given next inside <**script**> block in *index.html*.
- **Step 4**: Repeat the above three steps for each snippet given next.

3.10.3 Explore

Using any editor, we will write and test JavaScript code to use loops and see results on the web page.

In a nutshell, Objects can be looked upon as user-defined variables, which can hold multiple attributes and can be created with constructors as well to create Objects of similar types. Objects are mutable.

3.10.3.1 Code Snippet-1

```
// Utility ()
const appendToHTML = (contents, color = "white") => {
    contents.forEach((content) => {
        const myPara = document.createElement("p");
        myPara.style.color = color;
        myPara.innerHTML = content;
        resultHere.appendChild(myPara);
        sep();
    });
};
const sep = () => {
    const separator = document.createElement("hr");
    resultHere.appendChild(separator);
};
/***************************************************
********** */
var resultHere = document.querySelector(".results");

// Defining an object
var NetFlix = {
    content: ["Sherlock Holmes", "Enola Holmes"],
    customer: "Dibyasom Puhan",
};

// Printing the created object into webpage.
// Using Object.values(obj) to display it returns an
array.
appendToHTML(["Dummy object 1: ", Object.values(NetFlix)]);

// Using new keyword to create and define an object
var NetFlix1 = new Object();
NetFlix1.content = ["Sherlock Holmes", "Enola Holmes"];
NetFlix1.customer = "Ravi";

// Using JSON.stringify() to display the object.
// This method produces more readable information.
appendToHTML(["Dummy object 2: ", JSON.stringify
(NetFlix1)]);

// Defining object methods. ***************************
******
var NetFlix2 = {
    content: ["Sherlock Holmes", "Enola Holmes"],
```

```
        customer: "You",
        streamService: function (choice) {
            // Method definition
            return this.content[choice]; // This basically
refers to the current object being called, to resolve ambi-
guity, if any.
        },
    };
    appendToHTML(["Dummy object 3: ",
JSON.stringify(NetFlix2)]);

    //Calling Object method.
    const myStream = NetFlix2.streamService(1);
    appendToHTML([`Requested Stream: ${myStream}`], "tomato");
```

3.10.3.2 Code Snippet-2

```
    // Utility ()
    const appendToHTML = (contents, color = "white") => {
        contents.forEach((content) => {
            const myPara = document.createElement("p");
            myPara.style.color = color;
            myPara.innerHTML = content;
            resultHere.appendChild(myPara);
            sep();
        });
    };
    const sep = () => {
        const separator = document.createElement("hr");
        resultHere.appendChild(separator);
    };
    /********************************************************
******** */
    var resultHere = document.querySelector(".results");

    // Defining getter/setter to get and assign new value to
object properties.
    var NetFlix = {
        content: ["Sherlock Holmes", "Enola Holmes"],
        customer: "You",
        // Object Method
        streamService: function (choice) {
            // Method definition
            return this.content[choice];
```

```
            // It refers to the current object being called, to
resolve ambiguity, if any.
        },
        // Getter for customer attribute.
        get cust() {
            return this.customer.toUpperCase();
        },
        // Setter for customer attribute.
        set cust(newCust) {
            // Can check conditionals and then assign.
            if (newCust!== null) {
                // Or use any custom validity test.
                this.customer = newCust;
            } else {
                return;
            }
        },
    };
    appendToHTML(["Dummy    object   2:    ",    JSON.stringify
(NetFlix)], "yellow");
    // Using getter.
    const customer = NetFlix.cust;
    appendToHTML([`Current customer (Extracted via getter):
${customer}`]);

    //Setting new value to customer attribute.
    NetFlix.cust = "Dibyasom";
    appendToHTML(
        ["Setter   used,   dummy   object   current   value:   ",
JSON.stringify(NetFlix)],
        "black"
    );
```

3.10.3.3 Code Snippet-3

```
    // Utility ()
    const appendToHTML = (contents, color = "white") => {
        contents.forEach((content) => {
            const myPara = document.createElement("p");
            myPara.style.color = color;
            myPara.innerHTML = content;
            resultHere.appendChild(myPara);
            sep();
        });
    };
```

```
const sep = () => {
    const separator = document.createElement("hr");
    resultHere.appendChild(separator);
};
/*****************************************************
******** */
var resultHere = document.querySelector(".results");

// Defining blueprint to create instances i.e. objects of
same type
function NetFlix(content, customer) {
    // Object constructor
    this.content = content;
    this.customer = customer;

    // Object Method
    this.streamService = (choice) => {

        // Method definition
        if (choice < this.content.length) {
            return this.content[choice];
            // It refers to the current object being
called, to resolve ambiguity, if any.
        }
        else {
            return "Not Subscribed, sorry!";
        }
    };

    // Getter for customer attribute. | Can manipulate and
return better quality data.
    this.getCust = () => {
        return this.customer.toUpperCase();
    };

    // Setter for customer attribute.
    this.setCust = (newCust) => {
        // Can check conditionals and then assign.
        if (newCust !== null) {
            // Or use any custom validity test.
            this.customer = newCust;
        }
        else {
            return;
        }
```

```
        };
    }

    // Creating object with constructor.
    var nfAsia = new NetFlix(["Witcher", "Two man and a half"],
    "Dummy Dumdum");
        // Adding attribute to object constructor (via Prototype)
        NetFlix.prototype.region = "Global";

    var nfUk = new NetFlix(["Peaky Blinders", "Enola Holmes"],
    "Dummy Dumdumdum");

        // Printing newly created object.
        appendToHTML(
            [
                `Dummy object 1: ${JSON.stringify(nfAsia)}`,
                `Currently Streaming content-2:
    ${nfAsia.streamService(1)}`,
                `Currently Streaming content-6:
    ${nfAsia.streamService(5)}`,
            ],
            "black"
        );
        sep();
        appendToHTML(
            [
                `Dummy object 2: ${JSON.stringify(nfUk)}`,
                `Currently Streaming content-2:
    ${nfUk.streamService(1)}`,
                `Currently Streaming content-100:
    ${nfUk.streamService(99)}`,
                `Stream Service Location: ${nfUk.region}`,
            ],
            "yellow"
        );
```

4

Functions

Happiness should be a function without any parameters.

—Pranshu Midha

Functions are the first-class citizens of JavaScript. They are one of the important components while working with the language as they allow us to bundle up a set of working code. Functions are also objects in JavaScript, and they can be passed just like any other object. Functions help in building modular program.

This chapter explains the utility of Function object provided by JavaScript and further digs deep down into different ways to declare a function, learning about a new kind of arrow functions supported by JavaScript and explaining the concept of scoping, binding the variables with function. Functions are really important to understand in JavaScript as they provide us with a means to call an organized piece of code, which can be used repeatedly for different arguments.

4.1 Functions in JavaScript

A *function* defines a set of statement(s) or procedure(s) that can be executed by simply calling the function name. Functions also work like objects in a key value manner where the key represents the function name and the value represents the block of code that is associated with that name. It helps us to write elegant code by giving names and parameters to a set of statements that have a defined purpose. Avoiding code duplication is also one of the key reasons to use functions. JavaScript has some in-built functions like *alert()*[76], *confirm()*[77] and *prompt()*[78], which we have already discussed in previous chapters.

JavaScript allows us to generate user-defined functions dedicated to serving as a handle for execution for a set of statements. In JavaScript, anything other than primitive data types is considered an object. Function is a special object in the JavaScript language. This can be observed by executing the code *(function(){ }).constructor === Function*, which returns *true* [79]. Function as an object contains three properties, i.e. *length*, *name* and *prototype*. Further, *Function.prototype* contains many other properties and

DOI: 10.1201/9781003122364-4

methods, and some of them are explained here. Whenever we create a *function* in JavaScript, the created function will be an object of *Function* special object or blueprint; this can be checked using the *typeof* operator. Please note that Function is a special object, and function will be the user-defined object created using *Function.prototype*. To see the properties and methods of Function special object, type *Object.getOwnPropertyDescriptors (Function)* in the web browser's console [80].

Let us understand about the properties and few important methods available inside the *Function* object.

- *Function.prototype.name:* This property stores the name of the function defined; this can be checked using *functionIdentifier.name* inside the web browser's console.

- *Function.prototype.length:* This property stores number of arguments passed to a function; this can be checked using *functionIdentifier. length* inside the web browser's console.

- *Function.prototype.constructor():* This is a special method that creates Function object using the *new* operator. For example, *var myFunc = new Function().*

- *Function.prototype.toString():* This method is overridden from the Object prototype; this will print the code of function when used like *functionIdentifier.toString()* in the web browser's console.

- *Function.prototype.apply()/bind()/call():* These three methods are used for scope handling; these are discussed separately in the next section on function scope.

- *Function.prototype.arguments:* This property is an array-like object inside every function; this contains all the arguments passed to the function and facilitates access to arguments using index.

4.2 Function Variable Scope

Understanding the scope of a function and its binding is a little bit early to be discussed as of now. However, it is needed for you to understand it before moving ahead to function declaration.

The variables used in a function can be defined either outside the function body or within the function body. The variables defined within the function body are called *local variables.* Local variables are accessible to the associated function only. Local variables are initialized when a function is called and starts execution; when the function body completes execution, the variables are deleted. Such variables are only recognized within the function; therefore, variables with the same name can be used *outside* the function body.

The variables that are defined outside the function body are called *global variables*. These variables are accessible to any function that is defined throughout the code. If there is a local and a global variable with the same name, then the local variable shadows the outer one, meaning that the local variable is closely attached to its associated function and will cover the global value.

If there is no local variable available to the function, then the value of global variable is used for execution. Let us understand the scope of variables using the example given next:

```
// The following variables are defined in the global scope
var num1 = 20;
var num2 = 3;
var name = "Tom";
function multiply() {          // function is defined in the
global scope
    return num1 * num2; //can access global variables
}
// A nested function example
function getScore() {
    var num1 = 2, num2 = 3;
    function add() {
        return name + ' scored ' + (num1 + num2);
    }
    return add();
}
multiply();                          // Returns 60
getScore();                          //Returns "Tom scored 5"
```

The example demonstrates the use of global and local variables within a function. A nested function is also demonstrated; whenever *getScore()* is called, it will automatically execute *add()* and return the value.

That was about your understanding for scope of variables. Moving on toward the binding of variables within function calls. Just to have a little background, functions in JavaScript can be passed *as* objects within other functions, and whenever an object is passed to another function as a callback, the value of current object or *this* is lost. While developing web content, we need user-specific data to be passed to function every time it is called. Thus, losing the reference of *this* every time whenever a callback is performed leads to a problem. The solution to this problem is offered by function scope. There are three methods inside *Function.prototype*, i.e. *call()*, *apply()* and *bind()*. Let us understand their utility with the following example:

```
// Traditional Example
// A simple object with its very own "this" property
```

```
var dummyObject = {
    num: 100;
};
// Setting "num" on window to show how it is NOT used.
window.num = 2000
/* will discuss window in DOM chapter,
for now consider it a global scope */

// A traditional function to operate on "this"
var add = function (a, b, c) {
    return this.num + a + b + c;        //usage of this keyword
};
var result = add(1, 2, 3);
console.log(result)                              //prints   2006
(2000 + 1 + 2 + 3)
// using call
var result = add.call(dummyObject, 1, 2, 3);
// establishing the scope as " dummyObject "
console.log(result) // result 106 (100 + 1 + 2 + 3);

// using apply
const arr = [1, 2, 3];
var result = add.apply(dummyObject, arr);
// establishing the scope as "dummyObject"
console.log(result)          // result 106 (100 + 1 + 2 + 3)

// using bind
var result = add.bind(dummyObject)
//establishing the scope as "dummyObject"
console.log(result(1, 2, 3))                     // result 106
(100 + 1 + 2 + 3)
```

This example demonstrates how these three methods are used to associate scope of *dummyObject* while calling the method and while accessing the local *num* value. Whereas in the traditional function call, the global value of num, i.e. *window.num* is used. This basic understanding will help us in grasping the limitation of arrow functions discussed in upcoming sections.

4.3 Function Declaration

JavaScript supports numerous ways to declare any given function. In this section, we will discuss different approaches for declaring and defining a function.

4.3.1 Traditional Function Declaration

This method is a traditional way of declaring function that is supported since the very inception of JavaScript as a language. A traditional function is declared by using the keyword *function,* followed by the name of the function. Function names follow the same rules for naming that a variable does, i.e. alphabets, digits, underscore and dollar signs can be used. Please refer Section 2.2.5 for naming convention of variables. The syntax for defining a function, traditionally, is provided as follows:

```
Syntax:
function func_name (parameter1, parameter2, … parameter)
{
        //set of statements – body of the function
        //optional return statement
}
```

The keyword *function* is used for declaration followed by the name and the parenthesis, which may or may not have parameters defined inside them. A function can be created with empty parenthesis, i.e. no arguments. Single or multiple parameters can also be passed as function arguments separated via a comma operator. Parameters serve as additional information for the function that can assist in the execution of the body of the function.

The set of statements to be executed using the parameters is written within the curly braces. This section is called the body of the function. A return statement is optional in the function definition. A function may choose to send a return value or simply execute the body of function without returning anything. Mostly users need some output value post execution of their code, so a return statement is used. It is the last statement in the body of the function. If a return statement is not the last statement, then the function will execute the code up till return statement and return the value. While returning the value, it will come out of scope of function, and therefore, the statements following return will never be executed. Therefore, it is important to place return as the last statement within the code.

For example:

```
function add(num1, num2) {
    return num1 + num2;
};
var result = add(2, 3);
console.log(result)                    //prints 5
```

4.3.2 Function Expressions

While the previous function declaration is syntactically a statement, functions can also be created by using a function expression. The function created is anonymous; they do not need to have a name and are assigned to a variable. As Functions are also objects, so it is perfectly valid to assign its reference value to a variable [81]. For example, the function square could have been defined as follows:

```
const square = function (number) {
   return number * number;
};
var result = square(4);              // result gets the value 16
```

Function expressions are convenient when passing a function as an argument to another function. In the next example, *map* function is defined syntactically, and another anonymous function is created and assigned to *cubeCalculator* constant. Further, observe that *cubeCalculator* is passed to the *map* function just like any other object.

```
function map(placeHolder, a) {
    let result = [];              // Create a new Array
    for (let i = 0; i != a.length; i + +)
        result[i] = placeHolder(a[i]);
    return result;
};
const cubeCalculator = function (x) {
    return x * x * x;
};
let numbers = [0, 1, 2, 5, 10];
let cube = map(cubeCalculator, numbers);
console.log(cube);                // prints [0, 1, 8, 125, 1000
```

The previous example can be executed using *Array.prototype.map()* method, which we will discuss in detail in the chapter covering Arrays (Chapter 5).

```
var f = function (x) {
    return x * x * x;
}
    let numbers = [0, 1, 2, 5, 10];
let cube = numbers.map(f);
console.log(cube);                //prints [0, 1, 8, 125, 1000]
```

4.3.3 Arrow Function Expressions

An arrow function expression is a shorthand to write anonymous function expression using the special arrow => operator. These types of functions have their own advantages and limitations. The arrow function will always return some value. The arrow function can have a compact body or a block body [82]. For compact body, only a single line expression is required with no need of return statement. For a block body {}, a return statement is needed as demonstrated in the example:

```
var func = x = > x * x; // concise body syntax, implied "return"
var func = (x, y) => { return x + y; };
          // with block body, explicit "return" needed
```

To build a basic understanding of arrow functions, let us understand their working through the example that demonstrates how to convert a traditional function to the arrow function.

```
// Traditional Function to calculate square of a
function (a) {
    return a * a;
};
// Steps to convert to Arrow Function

/* 1. Remove the word "function" and place arrow between the
argument and opening body bracket, this is a block arrow
function connoting the {} and return statement */
(a) => {
    return a * a;
};

/* 2. Remove the body brackets and word "return" -- the
return is implied, this is compact body style */
(a) => a + 100;
// 3. We can even remove parentheses from arguments.
a = > a + 100;
```

4.3.3.1 Limitations of Arrow Functions

The arrow function provides a very concise style for writing functions, but they also come with their own set of limitations. These are briefly pointed out as follows:

- Arrow functions cannot be used as methods because they do not have their own bindings to *this* and *super*.
- Arrow functions are not suitable for call, apply and bind methods; these methods are used for establishing scope.
- Arrow functions do not have arguments property like we have while using the traditional methods.
- Arrow functions cannot be used as a shorthand for constructors.

Let us understand the first two of these limitations using some examples. The next example demonstrates the value of *this* inside a traditional function and an arrow function. Here, the *demoObject* is created with three properties, *a, arrowFunction* and *traditionalFunction*, where the latter two are arrow function and traditional function, respectively. Further, when these methods are called through *demoObject* object, we can observe that traditional Function is able to print value of 10 using *this* reference. *arrowFunction*, on the other hand, prints a *window* object through the *this* reference. This is because the *arrowFunction* does not have any local scope and is always defined in a global scope, which is a *window* object while working with JavaScript. As for *window* object, do not worry; we will be discussing it in detail in the DOM chapter (Chapter 7).

```javascript
var demoObject = {
    a: 10,
    arrowFunction: () => console.log(this.a, this),
    tradationalFunction: function () {
        console.log(this.a, this);
    };
};
demoObject.arrowFunction();
// prints undefined, Window {...} (or the global object)
demoObject.traditionalFunction();              // prints 10,
Object {...}
```

The next example is linked to the *call, apply* and *bind* methods that we have discussed before and demonstrates that they are NOT suitable for Arrow functions. For the reason that Arrow functions establish *this* based on the scope where the Arrow function is defined, i.e. generally *window* object.

```javascript
// Arrow Example of NOT supporting call, apply and bind.
    // A simplistic object with its very own "this".
    var dummyObject = {
        num: 100;
    };
    // Setting "num" on window to show how it gets picked up.
    window.num = 2000;
    /* For now consider window object to exist in global
scope */
    // Arrow Function
    var add = (a, b, c) => this.num + a + b + c;
    // using call on arrow function
    console.log(add.call(dummyObject, 1, 2, 3));
// result 2026
    // using apply on arrow function
```

```
const arr = [1, 2, 3];
console.log(add.apply(dummyObject, arr));
// result 2026
// using bind on arrow function
const bound = add.bind(dummyObject);
console.log(bound(1, 2, 3))
// result 2026
```

From this example, this should be clear to you by now that *call*, *bind* and *apply* methods did not show any impact on method call and every time the *window.num* is used. This justifies the statement of arrow functions for not supporting the scope methods of *Function.prototype*.

4.4 Parameters in a Function

Parameters are a set of values passed to a function that might assist in function execution. A function can be declared with zero, one or multiple parameters. Up to 255 parameters can be accommodated within the function's declaration. Parameters are passed when a function is called by its name. These parameters become arguments when received by the function. The values are copied to the function arguments, and thus, any change in these arguments is not actually reflected in the original values. When parameters are passed by reference, the change is reflected in the variables. The arguments of a function are not limited to strings or numbers. You can even pass whole objects to a function as a parameter to act on.

4.4.1 Return Keyword in Functions

The *return* statements can be used to return information to the calling variable. They are optional in a function declaration. They can be placed anywhere within the code, but mostly they are written as last statement. When a function encounters a return statement, it stops execution and returns to the calling code. Multiple return statements in a function allow the function to return to any one of them, depending upon the code design. An empty *return* statement can also be used. It will return *undefined* to the calling code.

4.4.2 Invoking a Function

Declaring a function only specifies what a set of statements do and attaches them to a name. In order to actually execute those statements, a function has to be *called* or, in other words, *invoked* by its name. In the example given next, *square* function is called with the parameter of 5 as *square(5)*. The result is then stored in the variable *z*. A function should be present within the scope from where it is being called.

```
const square = function (n) {
    return n * n;
} //declaration of method as expression
console.log(square(5));
// square is called with parameter
```

4.4.3 Recursion

Recursion is the process of repeating a task. While solving a problem, we can break it up and execute some statements over and over again to solve the bigger problem at hand. When this is done repeatedly for a piece of code, it is called *recursion* [83]. Best-suited example to demonstrate recursion is calculating a factorial for a number. For example, if we have to calculate factorial of 5; this is a recursive problem, and the solution is simple. If we know the value of factorial for 4 and further for the factorial of 3 and 2, respectively, ultimately the factorial of 1 is known to be 1, and the bigger problem can be merged to compute the actual problem of calculating factorial of 5. This is demonstrated programmatically using the following example:

```
function factorial(n) {
    if ((n = == 0) || (n = == 1))       //using OR operator
        return 1;
    else
        return (n * factorial(n - 1));
    /* recursive call to itself until a termination con-
dition is met */
};
```

4.4.4 Closures

A closure is the process of binding the nested function to its parent function's local variables [84]. This enables the nested function to use the local variables even when the parent function has ended its execution [85]. In other words, a closure gives you access to an outer function's scope from an inner function.

In JavaScript, closures are created every time a function is created. The following example demonstrates the usage of closures. In this example, we have defined a function *greeting(x)* that takes a single argument *x* and returns a new function. The function it returns takes a single argument *y* and returns the concatenation of *x* and *y*. We created two function expressions, i.e. *sayHi* and *sayBye,* from the *greetings* function by passing in *Hi* and *Bye,* respectively. The two functions returned by *greetings()* function are having the same function body with different values for *x,* i.e. *Hi* and *Bye.* Here, *sayHi* and *sayBye* are the closures created from the function *greetings().*

```
function greeting(x) {
    return function (y) {
        return x + y;
    };
};
var sayHi = greeting("Hi");
var sayBye = greeting("Bye");

console.log(sayHi("Tom"));          // Hi Tom
console.log(sayBye("Tom"));         // Bye Tom
```

4.5 Exercise

4.5.1 Theory

   i. Differentiate between function declaration and the function expressions.

   ii. Differentiate between the use of attributes and properties in functions.

   iii. What do you understand by closures in JavaScript?

   iv. Why do you think arrow functions were introduced in JavaScript? Explain the use of arrow functions with suitable code.

   v. What is the difference between *bind(), call()* and *apply()* methods used while working with functions? Use suitable program code to differentiate among these.

4.5.2 True/False

   i. It is mandatory for a function to return a value in JavaScript.

   ii. Function names are optional for the functions that are defined using expressions.

   iii. All the user-created functions inherit from the Function prototype.

iv. A variable number of arguments can be passed to functions at runtime.

v. *length* property can be used to check the number of arguments declared by a function in its declaration.

vi. *Function()* constructor is used to create a new function every time it is called.

vii. Functions in JavaScript cannot be created dynamically at runtime.

viii. JavaScript does not allow nesting of functions.

4.5.3 Multiple-Choice Questions

i. When the _____ method is invoked with an object as an argument, it returns a new function.
   a. valueOf()
   b. bind()
   c. call()
   d. apply()

ii. Newly created functions inherit from the prototype. Which of the following does it inherit from?
   a. Object
   b. Function
   c. Number
   d. Both a and b

iii. The _____ is defined as the combination of a function and its scope where the function's variables are defined.
   a. binding
   b. closure
   c. recursion
   d. None of the above

iv. In order to call a function, the following can be used:
   a. Function literal
   b. Function calling
   c. Function declaration
   d. Using prototype

v. What is the output of the following code snippet?

```
var x = 9;
function outerFunction() {
    var x = 3;
    function innerFunction() {
        x + +;
        var x = 4;
        console.log(x);
    }
}
outerFunction();
```

    a. Prints '4'

    b. Generates error

    c. Prints 3

    d. Prints '6'

4.6 Demo and Hands-On for Functions

4.6.1 Objective

    i. Define a basic function.

    ii. Understand length property, bind, call, apply and toString methods.

    iii. Understand the concept and example of JavaScript closure.

4.6.2 Prerequisite

- **Step 1:** Visit https://www.routledge.com/9780367641429 and download zip file for *Ch4_Function_Template*.
- **Step 2:** Unzip the content and open *index.html* in any editor of your choice (example: Notepad or Visual Studio Code).
- **Step 3:** Insert code given next inside <script> block in *index.html*.
- **Step 4:** Repeat the above three steps for each snippet given next.

4.6.3 Explore

Using any editor, we will write and test JavaScript code to define functions.

Here's what you should know about Function. Function is a named piece of code that can be called and executed any number of times, once defined. It reduces line of code, makes it easier to debug and makes the code modular.

4.6.3.1 Code Snippet-1

```
// Utility ()
const appendToHTML = (contents, color = "white") => {
    contents.forEach((content) => {
        const myPara = document.createElement("p");
        myPara.style.color = color;
        myPara.innerHTML = content;
        resultHere.appendChild(myPara);
        sep();
    });
};
const sep = () => {
    const separator = document.createElement("hr");
    resultHere.appendChild(separator);
};

/*********************************************************
******** */
var resultHere = document.querySelector(".results");

// JS Function Bind
// Creating an employee object
const employee = {
    _name: "JavaScript",
    getName: function () {
        return this._name;
    },
};

appendToHTML(["~ Function.bind() ~"]);
const unboundGetName = employee.getName;
// The function gets invoked at the global scope
appendToHTML([`Unbound Get name > ${unboundGetName()}`],
(color = "orange"));
// expected output: undefined

// Separator here
sep();

const boundGetName = unboundGetName.bind(employee);
```

```
  appendToHTML([`Bound Get name > ${boundGetName()}`],
(color = "lime"));
  // expected output: 42

/*******************************************************
******* */

  appendToHTML([
      "Function.length | Returns number of parameters ex-
pected by the function",]);
  sep();

  appendToHTML(
      [
          `appendToHTML.length > ${appendToHTML.length}`,
          "It's 1 as the other arg is optional.",
      ],
      (color = "lime")
  );
  sep();
  /*******************************************************
******** */
  appendToHTML([
      "call() | Calls a function with a given this value, and
individual arguments, that's the difference b/w call and
apply, as apply can only take one array arg.",
  ]);

  function Product(name, price) {
      this.name = name;
      this.price = price;
  }

  function Food(name, price) {
      Product.call(this, name, price);
      this.category = "food";
  }

  appendToHTML(
      [
          `Instantiating Product > ${new Food("MacOS Big-
Sur", 1000).name}`,
          `Instantiating Food, which calls Product con-
structor > ${new Food("cheese", 5).name
          }`,
      ],
```

```
        (color = "lightgreen")
    );
    // expected output#1: "MacOS Big-Sur"
    // expected output#2: "cheese"
```

4.6.3.2 Code Snippet-2

```
    // Utility ()
    const appendToHTML = (contents, color = "white") => {
        contents.forEach((content) => {
            const myPara = document.createElement("p");
            myPara.style.color = color;
            myPara.innerHTML = content;
            resultHere.appendChild(myPara);
            sep();
        });
    };

    const sep = () => {
        const separator = document.createElement("hr");
        resultHere.appendChild(separator);
    };

    /****************************************************
    ******** */
    var resultHere = document.querySelector(".results");

    appendToHTML([
        "Function.toString | Returns the function definition
    (source code) formatted to String",
    ]);
    sep();

    appendToHTML(
        [
            `appendToHTML.toString() > ${appendToHTML.
    toString()}`,
            `sep.toString() > ${sep.toString()}`,
        ],
        (color = "lime")
);
    sep();

    /****************************************************
    ******** */
```

```
appendToHTML([
    "Arrow Function | An arrow function expression is a
compact alternative to a traditional function expression,",
]);

// Traditional Function
function randomBetween(a, b) {
    return Math.random() * (b - a) + a;
}
// Arrow () equivalent
const randomBw = (a, b) => Math.random() * (b - a) + a;

appendToHTML(
    [`randomBetween(1,100): ${randomBetween(1, 100)}`],
    (color = "lightblue")
);
appendToHTML(
    [
        `randomBw(1,100): ${randomBw(
            1,
            100
        )} (Ps. Equivalent of randomBetween but declared
with arrow way.)`,
    ],
    (color = "orange")
);

sep();
```

```
/**************************************************/
```

```
appendToHTML(["JavaScript Closure"]);
// Add is assigned a self-invoking function, which set
counter = 0 and
// returns a function expression, that now makes 'add' a
function.
// Further calls invoke add, now the inner function, and it
retains the
// value of counter, as it is present in its parent scope.
// This is called closure, which makes JS functions have
PRIVATE variables.
// This functionality cannot be achieved by making counter
global, because any other function can increment it as well.
// With closure, privacy is enforced, and only add can in-
crement and retain the counter.
```

```
var add = (function () {
    var counter = 0;
    return function () {
        counter + = 5;
        return counter;
    };
})();

var counter = 0;
function addNonClosure() {
    var counter = 0;
    counter + = 5;
    return counter;
}

add();
add();

addNonClosure();
addNonClosure();

appendToHTML(
    [
        `add() returned ${add()}`,
        `addNonClosure()   returned   ${addNonClosure()}
//Should have been 15`,
    ],
    (color = "lime")
);
```

5

Arrays

We face a wide array of threats, which means we have to have a wide array of capabilities.

—Mac Thornberry

We have understood the concepts of object and function in JavaScript; now is the time to move on to arrays, and if we tell you that array is a single variable that is used to store different elements, you might wonder as to "Why should I learn arrays when I can simply store multiple values inside an object?" The answer to that thought lies in the fact that *Arrays* are the special objects in JavaScript, providing several features along with behaving like a normal object.

An object is perceived as the representation of a real-world entity, whereas an array is a collection of elements of different types and provides several built-in properties as well as methods to access and mutate this list. This chapter is dedicated to make you understand the concepts of Arrays and their usage of the built-in properties and methods in your JavaScript program. Unlike its counterpart in certain other languages, JavaScript arrays are a joy to use – this is something they definitely got right.

5.1 Array

An array traditionally is a continuous allocation of memory of predefined length. Though in JavaScript they are different in many cases, a JavaScript Array is a special object with a unique constructor, literal syntax, and an additional set of properties as well as methods inherited from *Array.prototype* [86]. In JavaScript, array is a single variable that stores multiple elements of same or different types. The values are stored in a list-like ordering that can be accessed using a numerical index starting from 0. Arrays cannot use strings as element indexes (as in an associative array), but they must use integers. The length and type of arrays are not fixed and can be changed anytime. The storage of array is also sparse, and hence, data can be stored and accessed at noncontinuous locations.

A normal object can also store multiple values, but it doesn't offer special functions, the way arrays do in JavaScript. These methods provide flexibility and feasibility for accessing and mutating the elements belonging to the array. The methods such as sort, search, insert, delete etc. represent the real power of using arrays in JavaScript, and we can even define our own methods to enhance the power of Arrays. Please note that an Array is a special object, and *array* is an object of Array. *Confused?* When we write *Array*, this means we are talking about the special object (blueprint) that will create *array* object (instance) for us.

5.2 Properties of Array

As discussed already, Arrays are special objects and, hence, they contain some predefined properties and methods. This makes them easy and efficient to use in programs. Properties of an Array can be retrieved by using *Object.getOwnPropertyNames(Array)*. The Array object (Array class) contains following instance properties and static methods:

- *Array.name:* This is a property that returns the string value for *Array*. This property is just the placeholder for the name of object.
- *Array.isArray:* This is a static method that returns *Boolean* output after checking whether the object passed as a parameter is the object belonging to Array or not.
- *Array.from():* This is a static method that is used to return a new array from a predefined Array object or an iterable object.
- *Array.of(elements)*: This is a static method used to return a new Array object from variable number or type of arguments.
- *Array.prototype.length:* This is property that is inherited by all the objects of Array type; this property returns the number of values in the instance.
- *Array.prototype:* This property is the most significant as it contains all the special built-in methods along with their default object methods. It is inherited by all the objects of Array type. Using these built-in methods by newly created array objects is powerful, and they can be used as needed by the developer.

5.3 Declaring an Array

Array manipulation is so frequent that soon you will not have to remember these declarations. They will come on their own with practice. However, we will still discuss array declaration in detail. There are two ways for declaring an array; i.e. either by using the Array literal or by the constructor. The following subsection explains them with example.

5.3.1 Array Literal

The square brackets used in array declaration are also called array literal or array initializer '[ ]'. Using the *var* keyword, we can create the array named *array_name*, and the values for initialization are kept inside '[ ]'. Each value is separated by using the comma operator. This way of declaring arrays is widely used because of its simplicity, faster execution speed and ease of readability of code.

Syntax:
```
var array_name = [val1, val2, val3, val4 … valN];
```

For example, the following array *worker* contains names of five workers. These names are stored in an array with index values starting from zero. *length* of the array will be 4 (0,1,2,3,4), which equals a total of five elements in the array with *Alice* having key 0, *Tim* having key 2 and *Charlie* with key value = 4.

```
var worker = ["Alice", "Bob", "Tim", "Kim", "Charlie"];
```

JavaScript also allows storing different types of data in a single array including integers, strings, empty object, objects etc. Observe the heterogeneous type of data in the same array instance.

```
var worker_hetrogeneous = ["Alice", { "id":
101,"name":"Bob" }, "Tim", "Kim", 101];
```

5.3.2 Using the New Keyword and Array Constructor

A new array object can also be created by using the *new* keyword that creates an instance of an Array. Values in the array should then be provided explicitly. The index of array starts with zero, as already discussed, but this is not a restriction to enter values starting from index zero only. The array can therefore be easily populated according to the needs of the program. Generally, this approach is avoided as it adds unnecessary complexity to code. Adding values in a sequential manner is definitely a simpler, cleaner and better alternative.

Syntax:
```
        var array_name = new Array()
```

Let us see an example for the same:

```
var worker = new Array()
worker[0] = "Alice"
worker[1] = "Bob"
worker[2] = "Tim"
worker[3] = "Kim"
worker[4] = "Charlie"
worker[21] = "Kate"
//Observe index, not recommended but valid
  worker[40] ="Smith"
//Observe index, not recommended but valid
```

Array can also be declared by using the *new* keyword and providing the values in the constructor itself. In this case, the index value will start from zero by default.

```
Syntax:
      var array_name = new Array("val1", "val2",…, "valN")
```

Let us see an example for the same:

```
var worker = new Array("Alice", "Bob", "Tim", "Kim",
Charlie");
```

5.4 Accessing an Array

JavaScript arrays can be accessed by using the array name and index number wrapped inside the square brackets '[ ]'. Index of array elements, therefore, plays a significant role while accessing and manipulating arrays. As we have learned, arrays are numbered collections of data, and these numbers are called indexes. The first element always has an index value of 0, and the last element has an index value of *length-1*. An entire array can be accessed by simply using the name of array. Please note that array are special objects, but the (.) dot notation cannot be used to access the elements of array. Let us see all these cases in the example given next:

```
var worker = ["Alice", "Bob", "Tim", "Kim", "Charlie"];
console.log(worker[0]);            //prints Alice
console.log(worker[3]);            //prints Kim
console.log(worker[worker.length - 1]);
//accessing the last element, prints Charlie
```

```
console.log(worker);
//prints entire array: ["Alice", "Bob", "Tim", "Kim",
"Charlie"]
console.log(arr.0);              //syntax error
console.log(arr.[6]);           //returns undefined
```

Further, the elements of an array can be modified by simply reassigning their values. JavaScript arrays are objects, and therefore, you can add or remove properties from an array with ease.

Storing array elements in an ordered manner is a good practice and thus widely followed. However, JavaScript allows you to leave empty indexes. It allows storing different types of data in a single array including integers, strings, empty object, nested objects etc.

```
var worker_hetrogeneous = ["Alice", { "id" : 101, "name" :
"Bob" }, "Tim", "Kim", 101];
```

5.5 Built-In Methods inside Array

We have mentioned built-in methods multiple times till now. The built-in methods in Array object are defined inside *Array.prototype* property of Array special object. These methods can be seen by entering *Array.prototype* in the browser's console. The new array instance that a developer creates from the Array's blueprint will inherit all these built-in methods. We can even define our own methods as *Array.prototype.function()* and then use it anywhere in our program. As per need, existing methods present in Array object can be overloaded as well.

The built-in methods are classified into three categories for ease of understanding. These categories are accessor methods, mutator methods and iterator methods, all of which are explained in detail with examples in the following sections.

5.5.1 Accessor Methods

Accessor methods operate on a copy of the calling array. In other words, they do not alter the original arrays. Instead, a copy of the array is created, and the method performs its operations on that copy of array. All the accessor methods are defined inside *Array.prototype*, and these are inherited by the objects created via invocation of *new* operator. Therefore, they are available under all these newly created array objects.

- *concat()*: Merges two or more arrays and returns the merged array. Sometimes, *concat()* without parameters is used to simply clone an existing array. It can also take values as a parameter and append those values at the end of the calling array. The syntax for using *concat()* method is given as follows:

Syntax:
```
        array_identifier.concat(value1, value2 ... valueN);
```

array_identifier is the name of the calling array object, and the associated values are passed as parameter to the *concat()* method. Passed arguments, i.e. *value*, can be either an array object or a data value. The parameters are concatenated in the order in which they are passed to the *concat()* method. For example:

```
var numeral = [101, 102, 103];
var text = ["Alice", "Bob", "Tim"];
var aplhaNumeric = [4024, "AH-2045", "TY-876"]
var result1 = numeral.concat(text);
//concatenates numeral and text and stores to result1
var result2 = numeral.concat(text, aplhaNumeric);
//concatenates 3 arrays [numeral, text and aplhaNumeric]
var result3 = text.concat(5, 6, [7, 8]);
// concatenates values at end of text and store in result3
var nested = [["Tim", "Bob"], 234, [56, 78]];
// creating a nested array
var result4 = nested.concat(text);
//concatenates nested and text retains the references
console.log(result1);
//prints [101, 102, 103, "Alice", "Bob", "Tim"]
console.log(result2);
/* prints [101, 102, 103, "Alice", "Bob", "Tim", 4024,
"AH-2045", "TY-876"] */
console.log(result3);
//prints ["Alice", "Bob", "Tim", 5, 6, 7, 8]
console.log(result4);
//prints [Array(2), 234, Array(2), "Alice", "Bob", "Tim"]
```

- *includes():* This method helps in checking whether a certain value is present in an array or not. If the value is present, it returns true; else it returns false. It is case sensitive in nature.

Syntax:
```
        array_identifier.includes(value)
        array_identifier.includes(value, index)
```

If only, value(s) are passed, it searches the entire array starting from the beginning. If an index is also specified, then the search begins from the index provided. If the index value is equal to or greater than the length of the array, then false is returned. Let us understand with an example.

```
var arr = [11, 22, 33, 44, 55, 66];
arr.includes(44);              //returns true
arr.includes(23);              //returns false
arr.includes(11, 2)            //returns false
arr.includes(55, 4)            //returns true
```

- *indexOf():* This method returns the index of the element where it is first found in the array. If the given element is not found in the array, then it returns −1.

Syntax:
```
array_identifier.indexOf(element);
array_identifier.indexOf(element, index);
```

In this syntax representation, *array_identifier* is the array to be searched. The element and index is the value from where the search begins. The value of *index* is optional, and if it is not provided, then the *index* is taken as zero by default.

```
var arr = ["a", "b", "c", "d","e", "f", "g"];
arr.indexOf("a")                      //returns 0
arr.indexOf("b", 2)                   //returns -1
arr.indexOf("h")                      //returns -1
```

- *join():* This method is used to join or unite all the elements of an array and produce a single concatenated string. By default, the elements of the array are separated by commas in the produced output string. We can optionally provide a separator to override the default comma separator.

Syntax:
```
array_identifier.join(separator);
```

The default separator is comma operator, but if you want to specify a different separator, then that is used in between the elements of the array. It returns an empty string, if the array is empty. If any element in the array is undefined, null or empty, then it is converted and displayed as an empty string. Let us understand its usage through the following example.

```
var colors = ['Red', 'Green', 'Blue'];
colors.join();              //returns 'Red,Green,Blue'
colors.join('+');           //returns 'Red + Green + Blue'
colors.join(' ');           //returns 'Red Green Blue'
```

- *lastIndexOf():* This method is used to find the last index at which an element is found in an array. It searches the array from backward starting from last index till the zeroth index, which represents the first element. If the element is found, its index is returned. If the element is not found, then it returns -1 to the calling function.

Syntax:
```
array_identifier.lastIndexOf(element, start_index);
```

start_index specifies the position (index) to start the search from. It is an optional parameter.

```
var arr = ["a", "b", "c", "b", 'a"];
console.log(arr.lastIndexOf('a'));     //returns 4
console.log(arr.lastIndexOf('b'));     //returns 3
console.log(arr.lastIndexOf('g'));     //returns -1
```

- *toString():* This method converts an array into a string value comprising comma separated values, similar to the default *join()* method. It is commonly used to retrieve the contents of an array in a string.

Syntax:
```
array_identifier.toString();
```

Array.prototype overrides the *toString()* method of Object. This method is called automatically when an array is to be used in string concatenation or when an array is to be represented as a text value.

```
var name = ['Tim, 'Bob, 'Alice'];
var str = name.toString();
console.log(str);           //returns "Tim,Bob,Alice"
```

- *slice():* This method is used to extract a sequence out of an array. It works on a copy of the array and does not alter the original array. Rather, a new array is created that is the copy of original array. *slice ()* method executes on this array, and the resulting array is sent back as a return value.

Syntax:
```
array_identifier.slice(index);
array_identifier.slice(start,end);
```

If only the *index* value is specified, then *slice()* creates a sequence that starts from index and ends on the last element of array. If both *start* and *end* indexes are specified, then *slice()* creates a sequence starting from *start* index up till the *end index-1*. The return value is the array comprising the elements that are sliced out or extracted from the original array.

5.5.2 Mutator Methods

Unlike accessor methods, mutator methods operate on the original array and not on a copy of it. These methods can add, alter, truncate or delete the elements of the array. Once modified, you cannot undo these operations. Every operation is performed on the original array; therefore, the developer must exercise caution while using these.

- *copyWithin():* This method is used to copy the elements of the same array into another location without modifying its length.

Syntax:
```
array_identifier.copyWithin(target,start,end);
array_identifier.copyWithin(target, start);
```

- *fill():* This method changes all elements in an array to a static value, from a *start* index (default *0*) to an *end* index (default *array.length*). It can work with different parameters as shown next and returns the resultant modified array.

Syntax:
```
array_identifier.fill(value);
array_identifier.fill(value, start);
array_identifier.fill(value, start, end);
```

In the syntax, *array_identifier* represents the name of the calling array, and the *value* is the element that is filled in the array. *start* is the index value from where to begin filling, and *end* is the end index. If only the *value* is passed as an argument without any index, the entire array is filled with the element value. If *start* index is provided along with the value, then the filling of array starts from the specified index and goes till the end of the array. If both *start* and *end* index are specified, then the *value* is filled between the indexes. If *start* or *end* index is negative, then it is calculated as *start + length* and *end + length*, respectively, to generate meaningful indexes. Usually, it is used for reverse indexing of arrays and operating on them.

```
var colors = ['Red', 'Green', 'Blue'];
var numerals = [1, 2, 3, 4, 5];
colors.fill('Red');
//returns 'Red, Red, Red'
numerals.fill(6);
//returns [6,6,6,6,6]
numerals.fill(7, 2);
//returns [1,2,7,7,7]
numerals.fill(8, 3, 4);
//returns [1,2,3,8,5]
numerals.fill(9, -3, -1);
//returns [1,2,9,9,5] as start:5-3 = 2 and end:5-1 = 4
```

- *pop():* This method operates similar to the pop operation on a stack data structure. It removes the last element of the array and returns its value. The original array is changed as one element from the end is removed. Therefore, the array length property is decremented by 1. Every time *pop()* is executed on an array, it shortens the array by one element.

Syntax:
```
        array_identifier.pop()
```

The given syntax returns the element that is removed or popped from the array when *pop()* is called with *array_identifier* being the name of array. The return type of this method depends on the elements of the array. If the calling array is empty, then *pop()* returns *'undefined'*.

```
var colors = ['Red', 'Green', 'Blue'];
console.log(colors.length);
//returns 3
var color = colors.pop();
//pops and returns 'Blue'
console.log(color);
//prints 'Blue'
console.log(colors.length);
//returns 2
```

- *push():* This method is similar in execution to the *push* operation performed on a stack data structure. It works just opposite to *pop()*, and it is used to add elements at the end of an array. It increments the array length by 1 before adding a new element to the end of the array. The arguments passed are inserted at the end of the array. This method returns the length of modified array after adding the element(s).

Syntax:
```
        array_identifier.push(element0);
        array_identifier.push(element0, element1);
        array_identifier.push(element0, element1,…, elementN);
```

This works similar to the *pop()* operation, as is clear with the given syntax. Let us understand this syntax using the following example.

```
var colors = ['Red', 'Green', 'Blue'];
console.log(colors.length);    //returns 3
var color = colors.push("Orange");
//returns 4 to color
console.log(color)
//prints ['Red', 'Green', 'Blue, 'Orange']
console.log(colors.length);    //returns 4
```

- *reverse():* This method is used to reverse the order of elements belonging to an array, i.e. first element becomes the last element and so on. The following syntax is used for array reversal where *array_identifier* is the array that we want to reverse. After reversing the order of elements, the reference to the array is passed as a return value.

```
Syntax:
        array_identifier.reverse();
```

Let us understand this using a simple example.

```
var colors = ['Red', 'Green', 'Blue'];
colors.reverse();
console.log(colors);
//prints ['Blue', 'Green', 'Red']
```

- *shift():* This method is used to remove the first element from an array. It works just like the *pop()* method with the difference being that *pop()* removes the element from the back of the array, whereas *shift()* removes the method from the beginning of the array. It returns the value of the removed element as a return value. The element at the zeroth index is removed, and the indexes of the rest of the elements are adjusted according to the change, i.e. second elements becomes first element and so on.

```
Syntax:
        array_identifier.shift();
```

If the array represented by *array_identifier* is empty, then *undefined* is returned as a return value. Otherwise, the first value belonging to the calling array is removed and returned.

```
var colors = ['Red', 'Green', 'Blue'];
console.log(colors.length);              //returns 3
```

```
var color = colors.shift();              //returns 'Red'
console.log(color)            //prints ['Green','Blue']
console.log(colors.length);              //returns 2
```

- *unshift():* This method is used to add elements to the beginning of array. It works like the *push()* method; the difference is that *push()* method inserts the elements at the end of array, and *unshift()* method adds elements at the beginning of the calling array.

Syntax:
```
        array_identifier.unshift(val1, val2,… valN)
```

Where, *val1, val2, … valN* are the elements to be added at the beginning of the calling array. This method returns the modified length of the array after insertion of elements. The order in which elements are passed is maintained, and the rest of the elements are re-inserted in the same order with modified index values.

```
    var colors = ['Red', 'Green', 'Blue'];
    console.log(colors.length);              //returns 3
    var color = colors.unshift("Orange");
//returns 4
    console.log(color);            //prints
['Orange','Red','Green','Blue'];
    console.log(colors.length);              //returns 4
```

- *sort(): This* method is used to sort the elements of an array. The elements are converted to string and then compared with their associated sequence of 16-bit Unicode Transformation Format (UTF-16) code values. The result is a sorted array, and it is returned back. Time and space complexity of the sorting depends on the implementation of the array. The sort method is explained in more detail in a separate section. The syntax for its use is given next.

Syntax:
```
        array_identifier.sort()
```

- *splice():* This method is a more refined way to manipulate an array. It is used to either replace or remove the elements present in an array. It takes parameters that will be used to replace the existing ones.

Syntax:
```
        array_identifier.splice(start, count, item);
        array_identifier.splice(start, count, item1,… itemN);
```

As usual, *array_identifier* is the reference of calling array where start is the index at which replacing (or removal) of elements will begin. If *start* < *length*, then start is set as the length of the array. If *start* < *0*, then start value is taken as *start = length-start*. If *length-start* < *0*,then *start* is taken as *0*. The second parameter *count* is the number that tells the *splice()* method about the number of elements to remove (or replace) in the array. This parameter is optional, and when not specified, *splice()* goes on deleting the elements till the end of the array. *item1, item2, ... itemN* are elements that are used to replace the existing elements from the calling array. If not specified, then *splice()* deletes only the items and does not replace them with values. For *splice()* method, the *return* value is an array containing the deleted values. If no values are deleted, then an empty array is returned. If only one value is deleted, an array with one element is returned.

```
var colors = ['Red', 'Black', 'Green', 'Blue'];
var spliced = colors.splice(2, 0, 'Orange');
// orange inserted at index 2 and no element is deleted as
count = 0
console.log(colors);
//prints ['Red', 'Black', 'Orange', 'Green', 'Blue'];
console.log(spliced);
//returns [] as no element is deleted
var spliced = colors.splice(3, 2) //deleting 2 elements
from index = 3
console.log(colors);          prints     ['Red',    'Black',
'Orange']
console.log(spliced);      //returns ['Green', 'Blue']
var spliced = colors.splice(1, 2, 'Yellow', 'White');
//deleting 2 elements from index1 and inserting
2 elements
console.log(colors);
//returns ['Red', 'Yellow', 'White']
console.log(spliced);     //returns ['Black', 'Orange']
```

5.5.3 Iterator Methods

Mostly when we have a list-like array and we need to access elements from it, we would have to traverse the entire array to find or process those elements. The JavaScript's solution is to offer simple loops as a means to iterate over the arrays. Another solution is using *Array.prototype* that provides some very useful methods to iterate over elements for an array. Iteration makes it easier to access the elements within arrays. JavaScript provides multiple ways to iterate over arrays, using the fundamental loops and predefined methods, which are topics of discussion for this section. Let's dive into it.

5.5.3.1 Using Fundamental Loops

The loop statements discussed as control flow statements in Chapter 2 can be used to iterate over the elements of an array. It is simple, easy-to-use way for iteration. Let us understand this using the syntax and some examples.

5.5.3.1.1 For Loop

One of the most basic ways for iteration is to use *for* loop, as array starts from index 0, so initial value is set to zero, the termination condition can be derived using the *Array.length* property. Similarly, the loop can also be iterated in a reversed manner using appropriate values.

```
Syntax:
        for (let i = 0; i < array_indentifier.length; i + +)
        {
            // Execute your operations on array element
        }
```

5.5.3.1.2 While Loop

Another way is to use the *while* statement for iteration. Do not forget to specify the iteration counter inside the loop; otherwise, the loop will continue execution without stopping.

```
Syntax:
        while(i < array_identifier.length)
        {
            // Execute your operations on array element
            i + +;
        }
```

5.5.3.2 Predefined Iterator Methods

Apart from the fundamental loops, *Array.prototype* provides few other methods to iterate elements inside an array. These methods may be used inside fundamental loops, also or they can be used in an independent manner. Iterator methods are used when you want to iterate or loop over the elements of array. Generally, some function is specified, which is applied to the elements of the array.

- *entries():* The *entries()* method returns a new Array Iterator object that contains the key-value pairs for every index present in the array.

```
Syntax:
        array_identifier.entries();
```

Example:

```
var colors = ['Red', 'Black', 'Green', 'Blue'];
const iterator1 = colors.entries();
console.log(iterator1.next().value);
//prints Array [0, "Red"]
console.log(iterator1.next().value);
//prints Array [1, "Black"]
```

- *forEach:* This method can be used for looping over arrays and running different functions on array elements. *forEach* is used to execute the specified function for every element of the array. It is very useful while manipulating entire arrays.

Syntax:

```
array_identifier.forEach(myFunction);
function myFunction()
{       //function which is executed for every element
        //set of statements for manipulating array
properties
}
```

- *every():* This method is used to test conditions on every single element belonging to the array. It returns a Boolean value depending upon the result of execution. It returns *true*, if all the elements of the array satisfy the condition specified within the *every()* method. Otherwise, it returns *false* even if single property within the array fails to pass the specified condition. If an empty array is passed, then it returns *true*, no matter what the specified condition is.

Syntax:

```
array_identifier.every(custom_function);
```

array_identifier is the reference name of the calling array, and *custom_function* is the function that specifies the test conditions for array elements. The specified test *conditions* or the *function* is executed upon every element of the array in a sequential manner. For every element, the processing is done, and if it comes out to true, only then the next element is tested. If *every()* returns false for even a single element, then the rest of the elements are not tested and *false* is returned.

```
function isLessThan(element, index, array) {
    return element = < 20;
}
var arr = [11, 5, 8, 13, 19]

var x = arr.every(isLessThan(20));
// returns true
```

```
[11, 54, 18, 130, 190].every(isLessThan(20));
// returns false
```

- *map():* This method can be used to run a function for every element of an array and then return a new array with the results.

Syntax:
```
        array_identifier.map(custom_function);
```

Example:

```
function multiplyByTwo(element) {
    return element * 2;
};
var input = [1, 2, 3, 4, 5];
var result = input.map(multiplyByTwo());
console.log(result)      //prints [2, 4, 6, 8, 10];
```

- *filter():* This method is used to create a new array out of the filtered values from a given array. The elements are filtered on the basis of some condition, which is passed in the method as an argument. This argument can be a simple conditional statement or a method in itself, which defined a set of filters for array elements.

Syntax:
```
        array_identifier.filter(custom_function);
```

Where, *function* is specifying the test conditions and *element* is the element of the array which is being processed. *function* is tested upon the element of the array, and if it passes the test specified by *function*, then it is added to a new array. In the same manner, every element is tested, and the elements that pass the function are added to an array. Elements that fail to pass these conditions are discarded and not added to the new array. This resulting array is then returned.

```
function isLessThan(element, index, array) {
    return element < = 20;
};
var numerals = [11, 55, 8, 23, 9];
var result = numerals.filter(isLessThan(20));
console.log(result);              //returns [11,8,9]
var result = [1, 54, 8, 130, 19].filter(isLessThan(20));
console.log(result);              //returns [1,8,19]
```

- *find():* This method is used to return the first element that satisfies the condition passed in the method. If you need to find the index of the first element that satisfies the given condition, then use the method *findIndex()*. If the value is not found, then undefined is returned.

Syntax:
```
      array_identifier.find(custom_function);
      array_identifier.findIndex(custom_function);
```

- *reduce():* The reduce operation is used to execute an aggregation operation, which is needed to maintain a state. This method takes a reducer function as an argument and executes it for every element present within the array to produce a single aggregated value as the output. The reducer function needs four arguments to work properly, i.e. accumulator, element, index and source array. Accumulator (or *acc*) works like a global variable throughout the method, and it is used to store the output value. Rest of the three arguments, i.e. element, index and source array, are all discussed before in previous sections. After processing every element, the value of *acc* is updated to hold the recent most value. Syntax for using *reduce()* method is provided as:

Syntax:
```
      array_identifier.reduce(custom_function(acc,value,
index,arr))
```

```
      array_identifier.reduce(custom_function(acc,value,
index,arr),initValue)
```

Where *custom_function* represents the *reduce()* function to be executed on each element of the array. *acc* holds the output after processing each element, and after the last operation, its value is returned. The second parameter is an optional initValue. If *initValue* is passed, then *acc* takes on this value to begin with. If *initValue* is not specified, then it is taken as 0 by default, and thus, *acc* assumes the value of 0 to start with. *element* is the current element being processed, and index is the index of the element being processed. *array_identifier* is reference name of the array on which *reduce()* operates; both index and array are optional to pass as arguments.

```
      var numerals = [1, 2, 3, 4]
      var aggregatedResult = numerals.reduce(function (acc,
value, index, array) { return acc + value });
      //no initial value given so acc = 0 to begin
      console.log(aggregatedResult);
```

```
//prints 10 calculated as: acc = ((((0 + 1)+2)+3)+4)=10
aggregatedResult = numerals.reduce(function (acc,
value, index, array) { return acc + value; }, 20);
   // initial value given so acc = 20 to begin
   console.log(aggregatedResult);
//prints 30 calculated as: acc = ((((20 + 1)+2)+3)+4)=30
```

- *some():* This method is used to test a condition on an array, and even if one element returns true for the condition, then this method is passed true as a return value. It returns a Boolean value as return type. If it is applied to an empty array, then false is returned.

Syntax:
        array_identifier.some(custom_function);

Where *custom_function* specifies the testing conditions on the elements of array. This method can also take index and array as arguments; they are optional to specify. If the testing function is passed by at least one element, then *true* value is returned.

```
var numerals = [1, 3, 4, 21, 4];
function lessThan10(element, index, array) {
    return element < 10;
}
numerals.some(lessThan10);              // true
[13, 16, 19, 12, 15].some(lessThan10);         // false
```

5.6 Nesting and Multidimensional Arrays

Arrays can also contain arrays within them, i.e. arrays can exist as elements of a different array. Such arrays are called *multidimensional arrays* [87]. Elements can be added via *shift()* and *push()* methods to the front and back of the array, respectively. Elements can be removed by using the *unshift()* and *pop()* methods from the front and back of the array, respectively. These arrays can be traversed either by using their indexes or by iterating over them as we have learned before. An example that demonstrates usage of multidimensional arrays is given as follows:

```
var nestedNumerals = [[1, 2, 3], [4, 5, 6], [7, 8, 9]];
console.log(nestedNumerals[1][1]);       // returns 5
console.log(nestedNumerals[2][1]);       // returns 8
nestedNumerals.pop();    //removes [7,8,9] from the array
nestedNumerals.push([12, 24, 48]);
//adds [12,24,48] to the end
console.log(nestedNumerals);
    //prints [[1, 2, 3],[4, 5, 6],[7, 8, 9],[12,24,48]]
```

Following examples demonstrate the usage of nested loop to iterate over the elements of multidimensional arrays:

```
var studentDetails = [["Tom", 80, 6], ["Tim", 75, 7],
["Bob", 88, 9]];
console.log("Name, Percentage, Class");
for (var i = 0; i < 3; i + +) {
   console.log("/n");
   for (var j = 0; j < 3; j + +) {
      console.log(studentDetails[i][j]);
   }
}
 /*Output:
  Name, Percentage, Class
  Tom,80, 6
  Tim, 75, 7
  Bob, 88, 9
 */
```

5.7 Sorting

JavaScript provides the *sort()* method to sort the elements of an array. The array elements are compared by value, and the resulting array is sorted in an ascending order. In order to find the descending order of elements, *reverse()* can be used on an array.

sort() works by converting the elements into strings and then comparing their values. Therefore, it might not work well in cases where numeric arrays need sorting. The solution to this is to pass a comparing function in the *sort()* method. Comparing function can be used to generate any kind of sorting operation for array elements. It gives us the flexibility to modify array elements in any order that we like. Syntax for using *sort()* is fairly simple and is provided next:

Syntax:

```
array_identifier.sort();
array_identifier.sort(comparingFunction);
```

The *comparingFunction* should return an integer value, and based upon the value, following decisions are made:

- *IfcomparingFunction (x, y) < 0*
  Then the value of x will come before the value of y in the sorted array.

- *IfcomparingFunction (x, y) > 0*
  Then the value of y will come before the value of x in the sorted array.

- *IfcomparingFunction (x, y) = 0*
  Then the order of the elements within the array remains unchanged.

The comparing function can be used to define different alternatives for sorting the array. Different operations can be performed on the elements while comparing and sorting. However, the sorted array will contain the same elements, i.e. operations carried out in comparing function have no effect on the value of elements present within the array.

The next example demonstrates usage of *sort()* and *reverse()* method on simple arrays and also its usage for comparing function for an array that comprises Objects. Observe carefully that the *customObjectArr* array object contains three objects as elements within the array. You should understand by now that it is not possible for *sort()* method to sort this custom array. In this case, we need to provide the logic for sorting objects, and hence, we define a function that can sort these objects based upon the keys of the object. Observe carefully the logic used in the example; it will help you to build a strong understanding of all the topics discussed so far.

```
var numerals = [4, 8, 1, 9, 3, 2, 8];
var text = ["Alice", "Bob", "Tim", "Carol", "Timmy"];
var specialArr = [{ "501": "Bob" }, { "101": "Tim" },
{ "202": "Carol" }];
numerals.sort();             // [1, 2, 3, 4, 8, 8, 9]
text.sort();
// ["Alice", "Bob", "Carol", "Tim", "Timmy"]
numerals.reverse()           // [9, 8, 8, 4, 3, 2, 1]
specialArr.sort(function (x, y) { return Object.keys(x)
[0] - Object.keys(y)[0] });
console.log(specialArr);
  /* Expected output:
    {101: "Tim"}
```

```
{202: "Carol"}
{501: "Bob"}
*/
```

5.8 Points to Ponder

Before finishing our discussion on arrays, let us think about some of the pointers for using arrays. It will be helpful to revise, summarize and elaborate on some critical points to supplement the rest of the discussion.

- An array does not have any predefined data type. It can store different data type values in a single array such as strings, numbers, Boolean and objects.
- Elements of an array can be easily accessed by using the index that is numeric in type.
- Array indexes always start from zero where [0] is used as an index value for the first element present in the array.
- The operator *typeof* returns the type *object* when it is applied on an object. For arrays, it returns Array if the object is created using the *new* keyword.
- An array can contain objects or other arrays or even functions as its elements. JavaScript is very flexible in such aspects.
- Different control flow statements can be used to iterate through the arrays.
- Objects use named indexes to store different values, whereas arrays use numbered indexes to store different values. That's why they are called as a special type of object. Numbered index storage makes them an ideal choice for storing large amounts of data. It is easier to work with for large-scale, ordered collections of data.
- JavaScript arrays come loaded with so many methods to play with arrays, thereby making it a popular choice among developers. As compared with other languages, arrays in JavaScript are more powerful and give a wide range of operations that can be executed on them.
- Arrays possess a global scope, and therefore, they can be easily accessed or manipulated throughout the life of the program.
- An entire array can be cleared or deleted if the length property is set to zero. Resetting the length property truncates and makes the array empty.

5.9 Exercise

5.9.1 Theory

i. Critically analyze the advantages and disadvantages of using arrays in JavaScript. Ideally, when should a developer use array in its program?

ii. What are the default values for different data types stored in an array?

iii. How can we compare, sort and slice an array? Use program code to demonstrate the working of these operations.

iv. Write a script to create three arrays using different ways for creation. Compare these arrays with one another, and then print the result on the console.

v. Write the program code for array addition, subtraction and multiplication of multidimensional arrays.

5.9.2 True/False

i. An array is a special object in JavaScript that is used to store same data types and values.

ii. An array can be directly used and populated without declaring it.

iii. *length* property can be used to retrieve the number of elements present inside an array.

iv. If an array has nine elements, then first element has the index zero and second last element has the index value as 7.

v. JavaScript arrays can be nested to create multidimensional arrays.

vi. Using *reverse()* method followed by the *join()* method will reverse the array and store it in the same array.

vii. Array size can be a negative number.

viii. Reachability is the central concept around memory management in JavaScript.

5.9.3 Multiple-Choice Questions

i. Using *pop()* method on an array will result in:

a. decrement of length by 1.

b. increment of length by 1.

c. no effect on array length.

d. None of the above

ii. For the following command, what will be the value of the fourth element?

```
var newArray = [1,3,4, ,8];
```

   a. Undefined

   b. Results in an error

   c. Zero

   d. None of the above

iii. For the following code snippet, chose the correct output:

```
var arr = ['abc', 5, 8, 11,'def', 6, 0,'ghi'];
arr.slice(0, 5);
```

   a. 'abc',5,8,11,'def'

   b. 'def',6,0,'ghi'

   c. 5,8,11,6,0

   d. Throws error

iv. An array can be identified using the _____ operator.

   a. ==

   b. ===

   c. typeof

   d. None of the above

v. Which of the following commands will you use to delete the array elements from a given array?
i) delete ii) shift() iii) splice() iv) pop()

   a. i and ii only

   b. iii and iv only

   c. None of the above

   d. All of the above

vi. Which of the following commands will you use to convert the array's length property into a read-only property?

   a. Object.defineProperty(arr, "length", {readonly:true});

   b. Object.defineProperty(arr, "length", {writable:false});

   c. Object.defineProperty(arr, {length:writable});

   d. Object.defineProperty(arr, "length", {readwrite:false});

5.10 Demo and Hands-On for Arrays

5.10.1 Objective

    i. CRUD actions on an array (Create, Read, Update, Delete).

    ii. Basic array utility operations.

    iii. Splitting a string into an array and coercing an array into a string.

5.10.2 Prerequisitev

- **Step 1:** Visit https://www.routledge.com/9780367641429 and download zip file for *Ch5_Arrays_Template*.
- **Step 2:** Unzip the content and open *index.html* in any editor of your choice (example: Notepad or Visual Studio Code).
- **Step 3**: Insert code given next inside <script> block in *index.html*.
- **Step 4**: Repeat the above three steps for each snippet given next.

5.10.3 Explore

Using any editor, we will write and test JavaScript code to interact with the DOM and see the results on the web page. Here's what you should know about Arrays. Arrays are collections of objects; they're heterogenous, meaning all the elements of an array don't have to be of the same type; and they are iterables and come with a vast variety of utility functions. Let's look into some important ones.

5.10.3.1 Code Snippet-1

```
// Reference to result field
const resultHere = document.querySelector(".result");

//Utility

// Let's create a few arrays.
const gotCharacters = ["Jhon Snow", "Hoddor", "Arya"];
const bigBangCharacters = ["Sheldon", "Raj", "Penny"];
const twoMenAndAHalfCharacters = ["Jake", "Alan",
"Charlie"];

// Parsing entire array as a string. Syntax:
array.toString()
// check browser console.
```

```
console.log("Original Array…");
console.log(bigBangCharacters);
console.log("Parsing as string!");
console.log(bigBangCharacters.toString());

const sep = () => {
    const separator = document.createElement("hr");
    resultHere.appendChild(separator);
};

// Utility function to append < p > </p > to HTML document
const appendToHTML = (contents) => {
    contents.forEach((content) => {
        const myPara = document.createElement("p");
        myPara.innerHTML = content;
        resultHere.appendChild(myPara);
        sep();
    });
};

appendToHTML([
    "Original Array: ",
    gotCharacters.toString(),
    "Pushing new element into array, result…",
]);

// Adding an element into an array.
gotCharacters.push("Sansa");
appendToHTML([gotCharacters.toString()]);

// Popping last element from array.
appendToHTML([
    "Original Array: ",
    bigBangCharacters.toString(),
    "Popping element from array, result…",
]);
bigBangCharacters.pop();
appendToHTML([bigBangCharacters.toString()]);
```

5.10.3.2 Code Snippet-2

```
// Reference to result field
const resultHere = document.querySelector(".result");
```

```javascript
//Utility

// Let's create a few arrays.
const gotCharacters = ["Jhon Snow", "Hoddor", "Arya"];
let bigBangCharacters = ["Sheldon", "Raj", "Penny"];
const twoMenAndAHalfCharacters = ["Jake", "Alan",
"Charlie"];

/* IMP, REST OF THE DOCUMENT USES ARRAY PARSED INTO STRING
FOR CONVENIENCE */
// Parsing entire array as a string. Syntax:
array.toString()
// check browser console.
console.log("Original Array …");
console.log(bigBangCharacters);
console.log("Parsing as string!");
console.log(bigBangCharacters.toString());

// Utility function to append separator
const sep = () => {
    const separator = document.createElement("hr");
    resultHere.appendChild(separator);
};

// Utility function to append < p > </p > to HTML document
const appendToHTML = (contents) => {
    contents.forEach((content) => {
        const myPara = document.createElement("p");
        myPara.innerHTML = content;
        resultHere.appendChild(myPara);
        sep();
    });
};

appendToHTML([
    "Array-1: ",
    gotCharacters.toString(),
    "Array-2",
    twoMenAndAHalfCharacters.toString(),
    "Concating these arrays, result…",
]);

// Concating two arrays into a new array.
const celestialArr =
gotCharacters.concat(twoMenAndAHalfCharacters);
```

```
appendToHTML([celestialArr.toString()]);
sep();

// Filtering Arrays.
appendToHTML([
    "Original Array: ",
    bigBangCharacters.toString(),
    "Filtering elements, word-length > 3… result > ",
]);
bigBangCharacters = bigBangCharacters.filter((element)
=> {
    return element.length > 3;
});
appendToHTML([bigBangCharacters.toString()]);
sep();

// Finding values. Returns the first value that matches
test, skips rest.
const naturalNumbers = [1, 2, 27, 4, 5, 6, 405404];
appendToHTML([
    "Original Array: ",
    naturalNumbers.toString(),
    "Finding a multiple of 3… result > ",
]);

// Function to check divisibility by 3.
const check3 = (num) => {
    return num % 3 = = 0;
};
appendToHTML([naturalNumbers.find(check3)]);
```

5.10.3.3 Code Snippet-3

```
// Create reference to button.
const btn = document.querySelector(".btn");

// Creating a reference to input box.
const myInput = document.querySelector("#thought");

// Create a reference to where result is to be displayed.
const resultHere = document.querySelector(".result p");

btn.addEventListener("click", () => {
    const processPrompt = document.createElement("p");
    processPrompt.innerHTML = "Doing this now…";
```

```
    processPrompt.style.color = "red";

    var content =

        "Splitting a string (your lyrics) into array of
string-chunks, based on a delimiter (white-space)";
    var comment = document.createElement("p");
    comment.innerHTML = content;

    resultHere.appendChild(processPrompt);
    resultHere.appendChild(comment);

    const lyrics = myInput.value;
    // Splitting the lyrics into separate words.
    const chunks = lyrics.split(" ");
    // Convert array into string, for displaying purpose.
    const result = document.createElement("p");
    result.innerHTML = `[ ${chunks.toString()}]`;
    // Display result.
    resultHere.appendChild(result);
});
```

6

Browser Object Model

I know what hierarchy and discipline mean. Without those, we will never have order and progress.

–Jair Bolsonaro

We have come a long way till now with all the discussion about the internals and fundamental concepts of JavaScript. We have also discussed objects, prototypes and special objects such as *functions* and *arrays*. This chapter will take us to some of the concepts related to real-life implementation of JavaScript. In this chapter, we will use all of the already-learnt concepts along with some special objects that are provided to us by the web browser itself. These objects enable us to make the behavior of website dynamic. This chapter is dedicated for understanding how JavaScript communicates with the web browser to perform certain operations. This is done by using the Browser Object Model (BOM). Let's dive straight into the definition and working of BOM without any delay.

6.1 Browser Object Model

BOM is a browser-specific convention used to expose all the objects associated with the web browser's window. This way, the program code can easily interact with the browser that is the host environment for the program and make changes to it as and when needed. There is no standard hierarchy for BOM, but all the major browsers today support it. For simplicity and consistency, we will consider the use of Google Chrome web browser throughout the discussion.

In simpler terms, JavaScript uses the BOM to interact with the browser's window. BOM comprises a hierarchy of objects that can be used to interact and modify the properties of the browser's window. To make JavaScript code interoperable on multiple platforms, nearly all the browsers support BOM and offer a universal *window* object that can be used to further access other objects provided by BOM. In tabbed browsers, every tab is represented by its own *window* object, and collectively, they all fall under the BOM hierarchy [88].

A reference hierarchy for BOM is depicted in Figure 6.1. The *window* object is present at the root of this hierarchy, and thus, every method or property automatically belongs to this object. By using these objects,

DOI: 10.1201/9781003122364-6

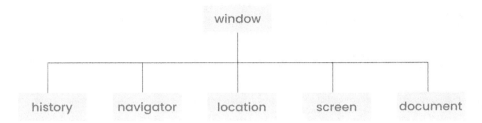

FIGURE 6.1
Reference hierarchy for BOM.

browsers offer many interfaces and in-built methods, which assist the developer to easily program and manage the interactions of program code with the browser. These objects have a specific set of methods that can be used across all the web browsers.

When web content is passed and loaded, BOM serves as the host environment, thereby providing functionality and assisting the task at hand, i.e. web content loading. Therefore, *window* object in BOM assumes a special role as a *global object*. Global nature of this object permits adding and modifying properties from the global scope. Although possible, it is not recommended to interfere too much in the global entities as it might disturb the hierarchy of objects and their attributes.

BOM hierarchy comprises the following objects:

- *window object*: It is the gateway to enter the BOM and serves as the root of BOM through which other objects provided in BOM can be accessed from the JavaScript code.
- *history object:* It is used as a store to history of a web browser in an array-like form; this object can be used to get/set history details.
- *navigator object*: It is used to handle and perform certain actions related to the navigation of web pages within a web browser, i.e. navigating a page forward or backward or reload.
- *location object*: It is used to access the URL of the loaded web page within a web browser.
- *screen object:* It is used to retrieve the screen size and other display-related properties from the web browser.
- *document object:* It is used to access the Document Object Model (DOM) for detailed markup structure of web page; in other words, this is the object that contains all the HTML + CSS + JavaScript code in the browser.

Figure 6.2 represents the object-wise mapping of BOM to the browser window. As depicted, *window* object represents the web browser's window.

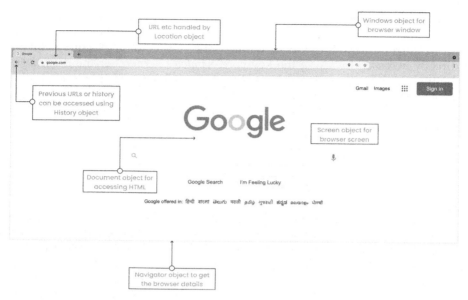

FIGURE 6.2
Visualizing objects in a browser window.

history object can be used to manipulate the URLs accessed by the user. *screen* object can be used to access the browser's screen-related properties. *navigator* object is used to access the details of the browser, and the *document* object is used to access, load and execute HTML code, which includes the JavaScript code. *document* object can be considered as the most important object existing in BOM. While all other objects deal with comparatively trivial information, *document* object helps to execute the code related to web content. DOM represents a separate hierarchy existing for *document* object; this will be discussed in more detail in the next chapter.

6.2 Window Object

In simple terms, the *window* object represents an open window in the web browser. The *window* object is a global object used to manipulate the browser window's life cycle and execute various operations within the browser window or tab [89]. It lies at the topmost level of BOM hierarchy and, thus, acts as the root object of a browser. It is an object belonging to the browser itself and not to the JavaScript code (JS objects include String, Array, Number etc.). *window* object represents the browser window itself and is supported by all the leading browsers. Therefore, all other objects, methods and properties down the hierarchy belong to the *window* object automatically.

The current browser window within which the code is executing can be directly accessed by using the keyword *window*. Remember the *cosole.log()* method, which we used to print output in the Chapter 1? It belongs to the *window* object, and every *window* object has its own *console.log()* method, which prints to the console of that particular web browser window. It is not mandatory to explicitly use the *window* reference while modifying the properties of the browser such as:

```
window.console.log("Hey, there!");
console.log("Hey there!");              //both are same
```

In the browser's console, simply writing *Reflect.ownKeys(window)* or simply typing *window* in the browser console will return the properties contained inside the *window* object. You will be surprised to see 1014 keys inside *window* object, which further have their own properties down the hierarchy. This demonstrates the breadth and depth of this tree structure containing properties for *window* object. This limits us to discuss all the properties and methods of *window* object here, and thus, we only discuss the most frequently used properties and methods in this chapter. Figure 6.3 represents the snapshot of console when *Reflect.ownKeys(window)* is executed.

For multitab browsers, every browser tab has its own *window* object. Whenever a new tab is opened, a new *window* object is created. Only the generic properties such as screen height, width etc. are shared between these objects, and thus, any change on them will affect the entire browser window.

In theory, *window* object is considered as the root of the tree whose branches are BOM, DOM, JavaScript, HTML and other front-end development codes. As represented in Figure 6.4, BOM can be considered as a group of objects that are made accessible to the developer through the *window* object.

FIGURE 6.3
Window object in browser's console.

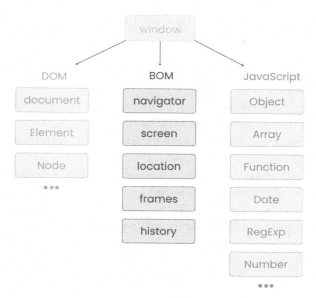

FIGURE 6.4
JavaScript interaction with BOM.

These browser windows/tabs can be further subdivided into *frames*. Frames are a unique way of dividing the browser window into sections where each frame can load its own script. Every frame object within a browser window is the child of the parent object. Frames are not universally supported by browsers, but they do provide a surgical approach to divide and operate on the sections of the browser window.

Window object defines specific methods and properties through which the browser window can be accessed and modified. These properties and methods are outlined as follows.

6.2.1 Properties of Window Object

window object has its own properties that are useful while interacting with this object. It is recommended to practice with these so that you imbibe these properties and use them when needed. These properties are explained as follows:

- *innerHeight and innerWidth:* It returns the inner height and width of the content area present in a browser window. It is useful for accessing the browser height and width for designing web content, which adjusts itself according to the user's browser specifications.
- *name:* This property returns the name of the current active window in a string format.

- *parent:* This property returns the parent object of the current window.

- *screenX and screenY:* This property returns the X and Y coordinates of the window with respect to the user's monitor/display resolution (in pixels).

- *top:* This property returns the reference to the topmost browser window that is open in the web browser.

- *status:* This property returns the text displayed on the status bar in a string format.

- *length:* This property returns the number of frame elements present inside the active browser window.

- *frames:* This property returns a reference to an array object representing the existing frames in the current window.

- *self:* This property returns a reference of the current window object belonging to the active/open browser window.

- *closed:* This property returns a Boolean value indicating whether the browser window is open or closed.

- *history:* This property returns the history object belonging to the *window* object, which can be used to access the URLs visited by the user.

- *navigator:* This property returns the navigator object of the window.

- *screen:* This property returns the screen object of the window that can be used to access width, height and other screen-related properties. *screen.height* and *screen.width* return the height and width of the user's browser window, respectively.

- *location:* This property returns the location object belonging to the *window* object. It can be used to access the current URL being visited by the user.

- *document:* This property returns the document object of the window, which can be used to manipulate the web content (HTML code) loaded by the browser. It represents every entity as an object within the browser, which can be modified to attain a desired behavior.

6.2.2 Methods of Window Object

The *window* object contains many methods that are used to interact with the end user and control the behavior of the browser window. The methods like *alert()*, *prompt()* and *confirm()* are few of the methods belonging to the *window* object. These methods provide dialog boxes that are used to interact with the end user. There are other methods like *setInterval()* and *setTimeout ()*; they support timing-based events. There are many other methods associated with the *window* object, some of which are explained as follows:

- *open():* This method will open a new browser window and loads the URL (if specified). It returns the reference of *window* object for the newly opened browser window. Parameters for *open()* are optional; if not specified, a blank new window is opened. The example of using *open()* is:

```
var url = "https://www.google.com";
var newBrowserWindow = window.open(url, name, width
height);
```

```
var newWindow = window.open(url, "New Window", "width =
300, height = 200");
/*open() returns an object reference to the variable
newWindow */
```

- *close():* This method closes the currently active window in the web browser.
- *print():* This method opens the Print Dialogue Box of the web browser window, which displays various print options for displaying content of the window. It has no return type.
- *scrollTo(width,height):* This method is used to scroll the document to the specified coordinates. It has no return type.
- *resizeTo(width,height):* This method is used to resize the browser window to the specified *width* and the *height* value.
- *resizeBy(width,height):* This method resizes the browser window by the given pixels, which is relative to the current size of the window.
- *blur() and focus(): These methods* are used to either remove or set the focus to the current browser window, respectively. *blur()* simply pushes the newly opened window to the background, and *focus()* brings it back to the front when needed. These are used to highlight key information to the user.
- *setInterval(callback_function, interval):* This method is used to execute a function at recurrent intervals (in milliseconds). This method continues executing at specified intervals unless *clearInterval()* method is called or the current browser window is closed using the *close()* method.

```
var x;
function newFunc() {
    x = setInterval(timeoutFunction, 2000);
    //runs the function every 2 second
}
function timeoutFunction() {
```

```
    console.log("I am a recurring function");
}
```

```
  newFunc();
//Calling function to start timed function
```

```
    window.clearInterval(x);
/* clears the interval and stops the execution, this should be
called when we want to stop recurring method*/
```

- *setTimeout(callback_function, interval):* This method is used to execute a function only once when the specified time interval is passed.

```
var x;
function newFunc() {
  x = setTimeout(timeoutFunction, 2000);
  //runs the function once after 2000 milliseconds
}
function timeoutFunction() {
  console.log("I am a recurring function");
}
newFunc();    //Calling function to start timeout function
```

- *stop():* This method is used to stop the loading of the current browser window.

There are many other methods, but all of them are not universally supported across browsers such as *createPopUp()* etc. Therefore, we selected the important ones to be discussed and remembered in this section. Moving on, we will now discuss other objects that are present within the *window* object.

6.3 History Object

History object is a property of the *window* object and contains the history-related information of the *session history* tab of browser [90]. Mostly, it is used to access an array of the URLs visited by the user. By manipulating the array, any page accessed by the user can be loaded on to the browser, including tabs and frames. It is simply accessed via *window.history* or *history* reference.

window.history is a read-only property, which is used to retrieve the reference to the *history* object. Every time a user clicks a link, a new entry is recorded in the browser's *history* object. These entries are stored in a stack-like structure, called *history stack*. This can be used to dynamically reconstruct web pages or to alter the web content loading behavior.

6.3.1 Properties of History Object

History object has the following properties associated with it. These properties can be used to manipulate the history of the browser.

- *length*: This property returns the length (in number) of URLs visited by the user during that web session.

  ```
  let numberOfURLVisited = window.history.length;
          //returns the length of array containing the URL's
  ```

- *current:* This property returns the current entry of the *history* object, i.e. URLs being visited by the user during that session.
- *next:* This property returns the next entry in the array of URLs visited by the user.
- *previous:* This property returns the URL of the last page visited by the user.
- *state:* This property contains a copy of all the entries belonging to the history object. This is mostly used for manually adding history entries via *pushState()* method. When a user enters a new state, a *popState()* event is fired. This event has a *state* property that holds the reference to current history entry.

6.3.2 Methods of History Object

The *history* object has the following methods that are used to take action on the accessed history of the user.

- *back():* This method loads the URL of previous page visited by the user. It works exactly the way when a user clicks the back button in browser window.
- *forward():* This method loads the URL of the next entry available in the array. In the same way, it works exactly the way forward button works in the browser window.
- *go():* This method is used to load a specific page from the session history array. It can be manipulated by using integers. For instance *history.go(-1)* will work like *back()* and *history.go(1)* will work like *forward()*. Similarly, any page can be accessed in any order via the *go ()* method, i.e. using -4 as a parameter to access the fourth last page visited by the user. Invoking *go()* without any argument will simply refresh the currently loaded web page.
- *pushState(state, title, URL):* This method is used to manually (via program code) add a new history event and manipulate the state of

history object. When the *pushState()* method is called with the specified URL, it is loaded and the top of URL stack is replaced by it, thereby changing the state of the history object. The URL must have the same origin as of the current URL; otherwise, an exception is thrown. Length property is incremented by 1 when *pushState()* method executes.

```
history.pushState(1, "T&F Homepage",
https://taylorandfrancis.com);
  history.state;                          //returns "1"
  history.pushState(2,"", '/about');
  history.state;                          //returns "2"
```

- *replaceState(state, title, URL)* replaces the current entry of the history object instead of adding a new entry. *state* variable holds the updated state and its corresponding title and loaded URL. *title* parameter is a string value representing title.

History object is used by developers to easily load and redirect to specific pages in order to smoothen the user experience. It is also useful while working with a single website with a lot of content. The developer can then use those windows and previously loaded URLs to serve the user. This enhances the user experience as the web page and URLs are not requested from the server but are loaded within the browser itself, provided they are loaded at least once by the user. Loading a URL that is completely different will throw an exception. This restricts the developer to load only those URLs whose parent is present and they have access to it. Not allowing any other URL to load is a safer alternative, which enhances security and decreases the chance of rogue code to manipulate the web page behavior in a malicious way.

6.4 Navigator Object

The *navigator* object is used to access information about the browser window such as the version, platform, operating system version, available plugins etc. The *navigator* object can be accessed via the *window* object, i.e. *window.navigator* or simply as *navigator*. It is a read-only property, and it can provide information about the host environment that is running the program code or script [91].

Navigator object is useful for certain specific cases, which makes overall website designing better. It can be used to check for compatibility of browsers and load specific scripts to resolve any compatibility issues. It can be used to access preferred languages of the user or to access location information to show custom content. It can also be used to check whether the browser is online or not via the *onLine* property. Information provided by *navigator* object can thus help in improving the end-user experience while navigating a website.

6.4.1 Properties and Methods of Navigator Object

The different properties belonging to the *navigator* object are briefly listed as follows. It is recommended for you to practice with these using your browser and script.

- *appName:* This property returns the name of the browser that is currently in use.
- *appCodeName:* This property returns the internal code name of the web browser.
- *appVersion:* This property returns the version number of the running browser as a DOM String.
- *cookieEnabled:* This property returns *true* if the use of cookies is enabled in the browser; otherwise, it returns *false*.
- *platform:* This property returns a string value containing information about the platform or machine type on which the web browser is running.
- *userAgent:* This property returns the *userAgent* String, i.e. header from server request for the currently running web browser.
- *plugins:* This property is used to return an array that provides information regarding the supported plugins for the web browser.
- *onLine:* This property returns the Boolean value *true* if the browser is online, i.e. it has a network connection; otherwise, it returns *false*.
- *language:* This property returns a string value representing the preferred browser language. The property *languages* return a String value comprising all the languages supported by the browser in order of user preference. This is useful while designing web pages for people who are geographically placed in different regions.
- *geolocation:* This property returns the *GeoLocation* object that holds the location information for the user's device.

Apart from aforementioned properties, there are many minimal methods available to operate on *navigator* object such as *javaEnabled()*; this returns *true* if Java support is enabled in the browser; otherwise, it returns *false*.

6.5 Location Object

The *location* object contains the information related to the currently loaded URL in the browser window. Both the *window* object and the *document* object can access the location object via *window.location* and *document.location*, respectively. *window.location* property stores the reference to the browser window's location object, whereas *document.location* stores the location object for a specified document [92]. It is important to understand the distinction between the *navigator* and the *location* object. *Navigator* object is used to access all the URLs, and it helps the developer to navigate through it, whereas *location* object is used to store detailed information only about the currently loaded URL within a web browser window.

6.5.1 Properties of Location Object

Location object offers many properties that can be used to query the address, host, port, protocol and other URL-related information. These properties are briefly explained as follows:

- *hash:* This property returns a String containing anchor portion (starts with #) of the URL.
- *href:* This property returns a String containing the loaded URL on the web browser window.
- *origin:* This property returns the protocol, host name and port number used by the currently loaded URL. It is a read-only property used to retrieve the origin of a specific location.
- *protocol:* This property returns a String containing the protocol scheme, which is needed to access that particular URL.
- *search:* This property is used to either set or return the search query for a URL in a String format.
- *port:* This property stores a string containing the port number used by the communication port of server for the associated URL.
- *pathname:* This property is used to either set or return the path name for the loaded URL.
- *host:* This property is used to either set or return a String containing hostname and port number for a URL, separated by a colon.
- *hostname:* This property is used either to set or return the host name or the domain name for the currently loaded URL in the web browser window.

6.5.2 Methods of Location Object

- *assign(URL):* This method is used to load a new document in the browser where URL is a String holding the URL for the document. It has no return value.

```
location.assign("https://www.google.co.in");
//loads a new URL
```

- *reload():* This method is used to reload the current document present in the *location.href* property in the browser. This method does not take any arguments and has no return type. It works equivalent to the reload button present in the browser.
- *replace():* This method is used to replace the currently loaded document with the specified document and replace the value of current history entry as well. This deactivates the *back* button making it impossible to return to the previously loaded document as it replaces the value of that URL itself.

```
location.replace("https://www.taylorandfrancis.com");
    //replaces the existing URL, removes the history entry
```

- *toString():* This method is used to retrieve a string containing the complete URL of the currently loaded document in the web browser. It is commonly used to retrieve the string value of the loaded URL.

6.6 Screen Object

The *screen* object holds the web browser's screen size–related information such as width, height, pixelDepth etc. It can be accessed by *window.screen* or via *screen* [93]. Properties associated with the screen object are outlined as follows.

6.6.1 Properties of Screen Object

- *width:* This property returns the total width of the current browser screen (in pixels).
- *height:* This property returns the total height of the current browser screen (in pixels).
- *availWidth:* This property returns the available width of screen, excluding the Windows taskbar space.

- *availHeight:* This property returns the currently available height of screen, excluding the Windows taskbar space.
- *pixelDepth:* This property is used to return the color resolution of web browser's screen (in bits per pixel).
- *colorDepth:* This property returns the color palette depth of screen (in bits). The number of bits represents the depth of color palette and is useful to know while displaying images.

Screen object has no standard method associated with it. However, its presence greatly helps the developer to design the web content while considering the user specifications of screen in mind.

6.7 Document Object

Document object is the core and the most important object contained within the BOM. It contains all the HTML elements present in a document that is loaded in a browser window. It simply represents the elements of the entire web page. Whenever an HTML document is loaded in a browser window, it becomes a *document* object belonging to the *window* object of BOM. In order to access any HTML element within the web page, one has to access the *document* object itself. It can be simply accessed via *window.document* or simply *document*. DOM is discussed in detail in the next chapter.

6.8 Exercise

6.8.1 Theory

i. How can you redirect a web page to load another web page on a given tab?
ii. What is the utility of different objects existing inside BOM? Explain each one individually with suitable attributes.
iii. Differentiate between *navigator* and *location* objects.

6.8.2 True/False

i. BOM is the de facto standard for JavaScript object model.
ii. BOM is supported in exactly the same way and is universally applicable to all the modern web browsers.

iii. BOM defines how the web browser interacts with the document's contents and with the host environment variables.

iv. *Navigator* and *location* objects have more or like same functionality; therefore, they can be used interchangeably.

v. While using *screen* object, the values used are in pixels.

6.8.3 Multiple-Choice Questions

i. The _____ object serves as the main entry point for all client-side JavaScript features.
   a. *navigator*
   b. *window*
   c. *history*
   d. *screen*

ii. If you want to execute an element in the browser repeatedly, which of the following methods would you use?
   a. *setTimeout()*
   b. *executeFunction()*
   c. *moveAnimation()*
   d. None of the above

iii. Which of the following objects serves as the root of BOM?
   a. *window*
   b. *document*
   c. *screen*
   d. *root*

iv. Objects within BOM are _____ in nature.
   a. global
   b. local
   c. program scope dependent
   d. Either global or local

v. A client's web browser window name can be accessed using _____.
   a. navigator.appName
   b. webBrowser.name
   c. client.navName
   d. None of the above

6.9 Demo and Hands-On for BOM

6.9.1 Objective

   i. Basic know-how of BOM.

   ii. Traversing forward/backward through browser session history.

   iii. Using *window* and other objects to manipulate the web page.

6.9.2 Prerequisites

- **Step 1**: Visit https://www.routledge.com/9780367641429 and download zip file for *Ch6_BOM_Template*.
- **Step 2**: Unzip the content and open *index.html* in any editor of your choice (example: Notepad or Visual Studio Code).
- **Step 3**: Insert code given next inside <**script**> block in *index.html*.
- **Step 4**: Repeat the above three steps for each snippet given next.

6.9.3 Explore

Using any editor, we will write and test JavaScript code to interact with the DOM and see the results on the web page. Here's what you should know about BOM. BOM stands for Browser Object Model. DOM is also a subset of it. It primarily allows JS to interact with browser-level elements and also provides a lot of information like screen size, hostname, protocol etc.

6.9.3.1 Code Snippet-1

```
// Utility ()
const appendToHTML = (contents, color = "white") => {
    contents.forEach((content) => {
        const myPara = document.createElement("p");
        myPara.style.color = color;
        myPara.innerHTML = content;
        resultHere.appendChild(myPara);
        sep();
    });
};
const sep = () => {
    const separator = document.createElement("hr");
    resultHere.appendChild(separator);
};
```

```
/*****************************************************
******** */
    var resultHere = document.querySelector(".results");
    // Check width allowed for rendering content (usable space).
    const windowWidth = window.innerWidth;
    const windowHeight = window.innerHeight;
    // Check actual resolution of the user's display.
    const screenWidth = window.screen.width;
    const screenHeight = window.screen.height;
    const windowLocObj = JSON.stringify(window.location);

    appendToHTML(
        [
            `Current window size: ${windowWidth} x
$ {windowHeight} (in Pixels) `,
            `Current screen size: ${screenWidth} x
$ {screenHeight} (in Pixels) `,
        ],
        "black"
    );
    sep();
    // Extracting information from window.location can be
useful in many cases, like enforcing or checking protocol,
hostname, etc...
    appendToHTML(["window.location holds these information
->", windowLocObj]);
```

6.9.3.2 Code Snippet-2

```
    // Utility ()
    const appendToHTML = (contents, color = "white") => {
        contents.forEach((content) => {
            const myPara = document.createElement("p");
            myPara.style.color = color;
            myPara.innerHTML = content;
            resultHere.appendChild(myPara);
            sep();
        });
    };
    const sep = () => {
        const separator = document.createElement("hr");
        resultHere.appendChild(separator);
    };
    /*****************************************************
******** */
```

```
    var resultHere = document.querySelector(".results");
    // Using window.history object to navigate in browser
history of the current session
    document.querySelector(".bwd").addEventListener
("click", (event) => {
        alert("I sense you want to navigate backward, granting
your wish (click OK)");
        window.history.back();
    });

    document.querySelector(".fwd").addEventListener
("click", (event) => {
        alert("I sense you want to navigate backward, granting
your wish (click OK)");
        window.history.forward();
    });
    appendToHTML([
        "In case you haven't done any browsing in the current
tab, buttons will have no effect.",
        "Copy this page's URL, and type something else like
RANDOM DOG PICS",
        "When new page opens, paste the copied URL, and hit
enter.",
        "And voila, click on the Backward button.", ]);
```

7

Document Object Model

In 1993 to 1994, every browser had its own flavor of HTML. So it was very difficult to know what you could put in a Web page and reliably have most of your readership see it.

—Tim Berners-Lee

This chapter discusses Document Object Model (DOM) in detail, as we have already learnt that DOM is a subpart of BOM, but DOM having its own significance is the reason this is discussed in a separate chapter. The web browser uses DOM as the interface for delivering content to the end user. DOM hierarchy represents the structure of XML and HTML documents being loaded inside the web browser. DOM represents the logical structure of documents, where a document refers to a web page. DOM is an object-oriented representation of the web page, which can be modified with a scripting language such as JavaScript. In this chapter, we will discuss the *document* object and its organization along with various concepts related to the DOM.

7.1 Document Object Model (DOM)

A web page is a document that comprises HTML. The web page can be viewed as a source HTML or inside a web browser. In both the cases, the document remains the same. The DOM defines a hierarchy of objects that serve as means to access, manipulate and manage the entire document [1]. Using this hierarchy, programs can easily get access to HTML tags, attributes, classes, IDs, elements etc.

The DOM implementation is done by web browsers, and most of the web browsers follow the standards [95]. The standard elucidates that any modern web browser on calling a method such as *querySelectorAll* should return all the elements that are specified as argument. Let's try this using a little bit of code, write this code in anyName.html file and open it in Chrome browser.

```
<html>

<head>
```

DOI: 10.1201/9781003122364-7

```
</head>

<body>
      <h1>This is The static Content</h1>
      <h2>The content below is created using JS and string is
taken input through prompt </h2>
      <h2>JS is fun with DOM!! </h2>
</body>

</html>
```

Now, open the console in the web browser, and type the following:

```
const heading2 = document.querySelectorAll("h2");
heading2;          //will print all nodes with h2
heading2[0];       //will show the inner content of this node
heading2[1];       //will highlight the selected node
```

The screenshot in the Figure 7.1 shows the output when the above code is executed on a web browser. Observe how different nodes and elements are

---

This is The static Content

The content below is created using JS and string is taken input through prompt

JS is fun with DOM!

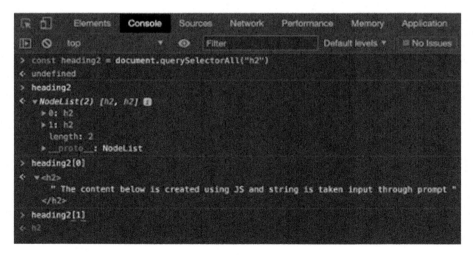

FIGURE 7.1
Output for document.querySelectorAll() property.

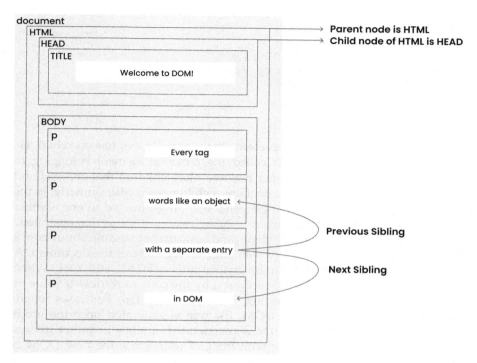

FIGURE 7.2
DOM representation of above HTML code snippet.

all accessible using the *document.querySelectorAll("h2")* method. The *document* object used in this example is the root for DOM. DOM comprises many such objects and nodes stored in a tree-like manner. These are discussed in detail in the upcoming sections.

Before moving ahead, let's see one more example to better understand the *document* object. In this example, a simple snippet of HTML code is given next that can be simply represented as a DOM structure as shown in Figure 7.2.

```
<html>

<head>
   <title>Welcome to DOM!</title>
</head>
```

```
<body>
    <p>Every tag</p>
    <p>works like an object</p>
    <p>with a separate entry</p>
    <p>in DOM</p>
</body>
<html>
```

In Figure 7.2, the *document* represents the root node that has one child, i.e. *<html>*. The element *<html>* is called the *document* element belonging to DOM. The *<html>* element further has two more children that are *head* and *body* elements. The *head* element contains a child named *title*; similarly, in the *body* element, there are different elements that are connected to one another using indexes and *previousSibling*, *nextSibling* relationships as demonstrated.

This example must have provided some basic understanding about how a DOM tree is generated for every web page on the basis of the document. A document is made up of markup languages such as HTML and XML. When DOM is manipulated, it is either accessed by the code to retrieve a value or is modified and the corresponding values are updated. For cases where modification happens via the DOM, the tree is generated once the modifications are done. When the tree is regenerated, it stops or rather freezes the interaction of web browser with the end user. Once the new DOM is generated, it is loaded on to the browser, and the end user is then allowed to interact with the tree.

DOM comprises different types of objects just like BOM. These objects can belong to XML, HTML, JavaScript or any other language that is used in manipulation of the document. It allows cross-platform compatibility and language independence across various browsers. The DOM along with JavaScript provides an application programming interface (API) that exposes the objects inside the document to be accessed and manipulated by the developer.

7.2 Relationship between BOM, DOM and JavaScript

We have already learnt that the BOM is created inside the special object *window,* which further has one of its children object called *document* object. This special object *document* is created by the web browser engine from the HTML or XML document that is loaded and contains a unique DOM structure inside it. The JavaScript code further can access these objects and allow dynamic modification to the browser as well as to the document [96]. This enhances the user experience and information delivery by mutating the BOM and DOM values. Entities within BOM and DOM tree are nearly all objects. These objects are accessed by the JavaScript code to generate dynamic behavior within a web page or a document. A document is a generic term used to identify HTML-based documents, XML-based documents, a web page and other information that is loaded onto a web browser for the end user.

BOM can be considered as a gateway to access the browser window and its associated properties, whereas DOM can be considered as the entry point to access all the objects belonging to a web page. JavaScript needs access to both BOM and DOM from time to time, depending upon the requirements. BOM is less frequently used to access the same properties again and again. DOM, on the other hand, is heavily used by JavaScript. DOM is the playground for JavaScript as it provides access to all the objects present inside a document that govern the structure of a web page.

To put everything in context, Figure 7.3 represents the positioning of BOM and DOM for a document. The *window* object is the root object of BOM, and we have learnt before that we do not need to add *window* before any of its child. So, we can use *document* object directly in the code that refers to the DOM. JavaScript can easily access the *window* object and *document* object–related entities using the root objects and their properties.

The *document* object is further branched into different objects depending upon the design and use of web page. As we can observe in the figure, *document* object has three other objects, i.e. *link*, *form* and *anchor*. The *form*

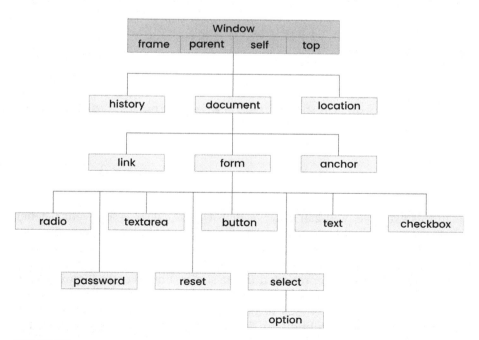

FIGURE 7.3
Window object and DOM.

object further has other children like *radio, textarea, text, checkbox, button* and many more. These objects also have child objects containing their own properties. This a highly variable tree structure as it changes for every document. Representation of DOM depends on the structure and type of document being loaded in the browser.

If the JavaScript code needs to access any of these objects, it can be done using the chained dot operator (.) over the *document* object. Every node in the DOM tree can be traced back to its parent. Let us understand this relationship using the example given next:

```
<html>

<head>
    <script>
        // run this function when the document is loaded
        window.onload = function () {
            var str = prompt("Enter any String")
            // create a h2 element inside HTML page
            const heading2 = document.createElement("h2");
            const heading2_text = document.createTextNode(str);
            heading.appendChild(heading_text2);
            document.body.appendChild(heading2);
        };
    </script>
</head>

<body>
    <h1> This is The static Content </h1>
    <h2> The content below is created using JS and string is
taken input through prompt </h2>

</body>

</html>
```

In this example, you can observe that *window.onload* is an event listener that is associated with a function. The function is taking input from the user using the *prompt* function and creating *h2* element for display to the user. Heading *h2* is appended in the HTML DOM using different methods. The output of this code is represented in the screenshot given in Figure 7.4.

In the following sections, we would be learning some of the frequently used properties and methods belonging to the *document* object that help in providing more flexibility to the developer for working around the web page.

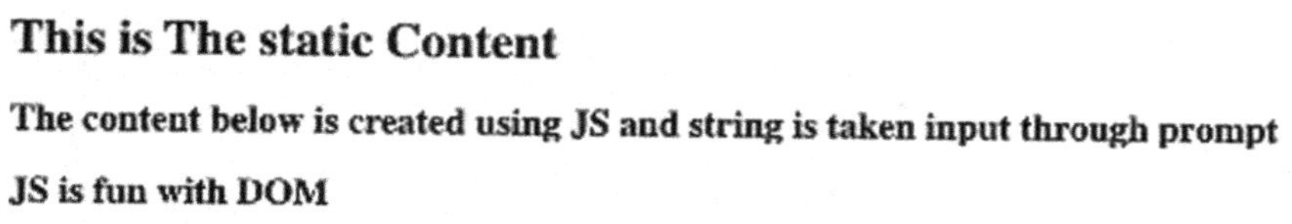

This is The static Content

The content below is created using JS and string is taken input through prompt

JS is fun with DOM

FIGURE 7.4
The output for above code.

7.3 Understanding DOM Tree and Nodes

DOM is quite often called *DOM tree* having objects as children that are called nodes. For every HTML document parsed by the web browser, a corresponding DOM tree is generated with a single root node called the *document* object and subsequent child nodes. Root node is the parent to the HTML element tag and is often called the *document element* or *object*. Every document has only one *document element*. There can be more than one branch arising out of *document* for child nodes, depending upon the complexity of the web page being loaded [97]. Before moving forward, let us first revisit some HTML terminology, and then we will create a DOM tree using the example given next:

```
<!DOCTYPE html>
<html lang = "en">

<head>
    <title>Lets Understand DOM </title>
</head>

<body>
    <h1>This is first example </h1>
    <a id = "home" href = "index.html">Home </a>
    <!-- Creating a href with id -->
</body>

</html>
```

For this example, there are different tags, attributes and other entities belonging to this program code. If one has to segregate these into categories, it can be done simply as follows:

- *Tag*: html, head, title, body, h1, a
- *Attribute*: lang, id, href
- *Attribute Value*: en, home, index.html

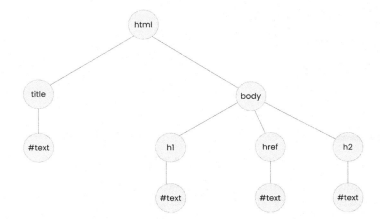

FIGURE 7.5
DOM tree representation.

- *Text*: Lets Understand DOM, This is first example, Home
- *Comment:* With the value 'Creating a href with id'.
- *Element:* Everything between opening and closing tags is called an element.

To visualize the DOM tree created from the previous HTML document, we used [98] to create Figure 7.5. Try to make some more demo tree structures to get a hold on it.

Each node in the DOM tree is identified using a node type. The type of a node can be examined using the property *nodeName.nodeType*. This will return an unsigned short number that refers to the constant defined inside *Node* property. Some of the node type constants and their associated values are given in Table 7.1.

TABLE 7.1

Node Type Constants

Value	Constant	Explanation
1	Node.ELEMENT_NODE	A simple element node such as <p><h1><div> etc.
3	Node.TEXT_NODE	Represents the *text* present in an Element or an Attribute
4	Node.CDATA_SECTION_NODE	Represents a CDATASection
8	Node.COMMENT_NODE	Represents a *comment* node
9	Node.DOCUMENT_NODE	Represents a *Document* node
10	Node.DOCUMENT_TYPE_NODE	Represents a *DocumentType node*, i.e. <!DOCTYPE html>

```
> document.nodeName
<- "#document"
> document.querySelector("#home").nodeType
<- 1
> document.querySelector("h1").childNodes[0].nodeType
<- 3
> document.querySelector("body").childNodes[4].nodeType
<- 8
```

FIGURE 7.6
Screenshot for using *document* object.

Out of multiple node values present, in our example, we have used Element node, Text node and Comment node. You check this by opening the browser's console and running the previous HTML code. Enter the following commands on the console. The output of using these commands on the console is represented via Figure 7.6.

```
document.nodeName
document.querySelector("#home").nodeType
document.querySelector("h1").childNodes[0].nodeType
document.querySelector("body").childNodes[4].nodeType
```

The screenshot (Figure 7.6) shows output received.

In our example, we can observe that various nodes are created within DOM, and we can easily classify these nodes into following three categories: Element Node, Text Node and Comment Node.

7.3.1 Properties of Node Object

For every node there are many properties associated with it, like in the example given previously that we have used the *nodeType* property [99]. Some of the other properties belonging to *Node* objects are explained as follows:

- *baseURI:* This property returns a String containing the base URL of document.
- *isConnected:* This property returns a Boolean value specifying whether the node is connected to the object or not, directly or in-directly. It returns *true* if the child is directly connected to the *document* object; else it returns *false*.

- *nodeName:* This property returns the name of the calling node in a String format.
- *nodeType:* This property returns an unsigned short value indicating the node type, as elaborated in Table 7.1.
- *firstChild* and *lastChild:* This property returns a *Node* object holding the first and the last child, respectively, for a given calling node.
- *childNodes:* This property returns a list containing all the child nodes belonging to the calling node.
- *parentNode:* This property returns the parent *Node* object for the calling node; otherwise, it returns *null.*
- *parentElement:* This property returns the parent *Element* object for the calling object; otherwise, it returns *null.*
- *nextSibling* and *previousSibling:* This property returns the next and previous sibling of the calling node, respectively. It returns *null* if no such node exists.
- *textContent:* This property is used to either retrieve or set the text content of the calling node and all its subsequent children.

7.3.2 Method of Node Object

The *node* object also contains many methods that are helpful in accessing and working around with the DOM nodes. These are further explained as follows:

- *contains():* This method returns a Boolean value as a result after checking if a node is a child of the calling node or not.
- *compareDocumentPosition():* This method is used to compare the position of one node with respect to another node, which is passed as an argument. This method returns an integer value that defines the corresponding output in constants that are defined inside Nodes object.
- *appendChild():* This method is used to append a new child node at the end of the calling node.
- *getRootNode():* This method returns the root object for the calling object.
- *insertBefore():* This method is used to insert a node just before the calling node.
- *cloneNode():* This method is used to clone or copy a node along with its contents.
- *isEqualNode():* This method returns *true* if two nodes hold the same contents including their data types. If not, then it returns *false.*
- *isSameNode():* This method returns *true* if the two nodes point to the

same object, i.e. calling node and passed node are same. If the nodes are not same, then *false* is returned.

- *removeChild()*: This method is used to remove a child node from the calling parent element.
- *replaceChild()*: This method is used to replace a given node with another node, which is passed as the parameter.

7.4 Document Object

The root of the DOM tree is the *document* object [100]. This object provides different properties and methods that provide access to different nodes in the HTML document. *document* object is created by the web browser from *Document* interface, which is inherited from Node interface. A *document* object can also be created dynamically using the constructor *Document()*. The *document* object can be directly used with or without using the *window* prefix. Most of the properties belonging to *document* are read-only in nature. The following sections explain frequently used properties and methods belonging to the *document* object.

7.4.1 Properties of Document Object

The *document* object contains many properties that help the developer to get metadata about the host environment such as character set used, head sections and much more. There are many properties in the *document* object, but we are discussing some of them here.

- *body*: This property returns the *<body>* node of DOM. If the body tag is not present in the source document, then it returns *null*.
- *activeElement:* This property returns the currently active Element object.
- *characterSet:* This property returns a String containing character encoding used by the document.
- *children:* This property returns an object containing all the child elements of the calling document.
- *childElementCount:* This property returns the total number of child elements present in a document.
- *doctype:* This property returns an object containing the Document Type Definition, i.e. DTD of the document.
- *documentElement:* This property is used to return the element that is the immediate child for the document.

- *document.URI:* This property is used to return the document's location in a string format.
- *embeds:* This property returns the list of all embedded elements *<embed>* present in the document.
- *forms:* This property returns the list of all the form elements *<form>* present in the document.
- *head:* This property returns the HTML *<head>* element.
- *images:* This property returns a list of all the images present in the document.
- *implementation:* This property returns the DOM implementation of the present document.
- *links:* This property returns a list of all the hyperlinks present in the document.
- *plugins:* This property returns a list of all the available plugins for the user's web browser.
- *scripts:* This property returns all the script elements *<script>* for the calling document object.
- *visibilityState:* This property is used to return a String that informs about visibility of document with possible values being *visible, hidden, unloaded* and *prerender.*

Apart from these properties, *document* object also inherits from *HTML Document* interface, which is applicable since HTML 5. Inheriting from this interface adds properties such as *cookie, designMode, dir, readyState, lastModified, defaultView, referrer, title, URL* etc.

7.4.2 Methods for Document Object

The methods belonging to the document object are of great utility, and these are used very intensively while developing a web page using JavaScript. These methods provide operations such as creation of a node, modification of nodes, deletion of nodes and querying the nodes. Among many methods, two of them deserve special attention here, i.e. *querySelector* and *querySelectorAll*. The *querySelector* method is used to return an element or a node matching the passed argument, and the *querySelectorAll* returns a list of all the elements matching the passed argument. The list of other frequently used methods for *document* object are given as follows:

- *append() and prepend():* These methods are used to insert objects at the end and beginning of the calling document, respectively.
- *adoptNode():* This method is used to adopt a node from an external document.

- *createAttribute():* This method creates a new *Attribute* object and returns it.
- *createComment():* This method returns a newly created *Comment* node.
- *createCDATAsection():* This method returns a newly created *CDATA* node.
- *createElement(name):* This method returns a newly created *element* node with the specified name.
- *createEvent():* This method returns a newly created *Event* object.
- *createTextNode():* This method returns a newly created *Text* node.
- *replaceChildren():* This method replaces the children of document with the new node(s), which are passed as arguments within the method.
- *querySelector():* This method returns the first matched element node in the document that matches the criteria.
- *querySelectorAll():* This method returns a list of all element nodes in the document that match the criteria.

7.5 Elements in DOM

When an HTML element is an item in the DOM, it is referred to as an element node. Element is primarily a base class from which every *element* object in the document inherits its functionality [101]. Every *element* object inherits from its parent *Element* class. These objects represent common HTML elements such as *<p></p>*, *<a></a>*, *<div></div>*, *<head></head>* etc. Anything between opening and closing tags is considered as an element in DOM.

It is often confusing to distinguish between *node* and *element*. Both of them are entities of DOM, but *node* is a generic term referring to any of the objects present inside DOM, whereas *elements* are specifically those nodes that have a numeric constant property *Node.ELEMENT_NODE* with a value 1 associated with them. Therefore, an element is basically a special type of a node. Every element with Node.ELEMENT_NODE = = 1 is a node, but every node is not an element.

The Element is also derived from Node, and hence, you will find many repeated properties as well as methods. However, these are provided again to avoid any confusion later on. The next sections explain these properties and methods belonging to the element object.

7.5.1 Properties of Element Object

The *element* object contains many properties that help to retrieve attributes, child and parents of the nodes. The frequently used properties are listed next:

- *attributes*: This property returns an object with its associated attributes for the calling element.
- *id*: This property returns a DOMString containing the unique global ID for the calling element.
- *localName:* This property returns the local name of calling element via a DOMString.
- *children*: This property returns a read-only *HTMLCollection* object with the child elements belonging to the calling element.
- *firstElementChild* and *lastElementChild:* This property returns the first and last child element belonging to the calling element.
- *childElementCount*: This property returns the number of children of calling element.
- *classList*: This property is used to retrieve a list of *class* attributes for the calling element.
- *className*: This property is used to either retrieve or set the class attribute value for the calling element.
- *clientHeight, clientLeft, clientTop* and *clientWidth*: This property returns a Number object with the inner height, left border, top border and inner width of element, respectively.
- *nextElementSibling* and *previousElementSibling*: This property returns the immediate next and previous siblings for the calling element, respectively.

7.5.2 Methods for Element Object

Similar to the *document* object methods, the methods belonging to *element* interface are inherited from the *Node* interface. Some of the frequently used methods are briefly explained in this section. Don't get surprised when you see *querySelector* and *querySelectorAll* here again. They belong to *element* object as well.

- *append()*: This method is used to insert a *Node* or *DOMString* object at the end of the calling element.
- *addEventListener()*: This method is used to add an event handler with an event type on the calling element.
- *animate()*: This method returns an animation object that can run animation on the calling element.
- *getAnimations()*: This method returns an array of currently active animation objects on the calling element.
- *getAttribute()*: This method returns the attribute belonging to the element in an object form.

- *hasAttribute()*: This method is used to check whether an attribute belongs to the element or not. It returns a Boolean value.
- *setAttribute()*: This method is used to set the specified attribute on the calling element.
- *prepend()*: This method is used to add an object (*node* or *DOMString*) before the first child of the calling element object.
- *querySelector()*: This method returns the first *element* or *node* object, which matches the specified criteria passed as arguments within parenthesis.
- *querySelectorAll()*: This method returns a *NodeList* object containing all the objects qualifying the specified criteria.
- *remove()*: This method is used to remove an element from its parent element.
- *removeAttribute()*: This method is used to remove the named attribute on the calling element.
- *replaceChildren()*: This method is used to replace child node(s) with the passed arguments that serve as nodes.
- *getElementByClassName()*: This method is used to return an *element* object whose *ClassName* property matches the passed string as the argument.
- *getElementByTagName()*: This method is used to return an *element* object whose *TagName* property matches the passed string as the argument.
- *getElementById()*: This method is used to return an *element* object whose ID property matches the passed string as the argument. ID property is a case-sensitive string value that is unique for every element.

The next section will exploit few of these functions to explain how to access and manipulate different elements present in the DOM structure.

7.6 Accessing Elements in the DOM

Till now we must be having a good understanding of *document, node* and *element* along with their properties and methods. This section is intended to elaborate how we can access the HTML elements present inside DOM [94]. To proceed forward, we should have elementary knowledge of some CSS selectors and HTML-like syntax such as defining *tags, class* and *name*. Along with these, we would be using following methods from DOM objects: *getElementById, getElementsByClassName, getElementsByClassName, querySelector* and *querySelectorAll.* To continue, type this code in any notepad editor, and save your file with *anyName.html.* The code is given in the next example:

```
<!DOCTYPE html>
<html lang = "en">

<head>
    <meta charset = "utf-8">
    <meta name = "viewport" content = "width = device-width,
initial-scale = 1.0">
    <title>Accessing Elements in the DOM </title>
    <style>
        html {
            font-family: sans-serif;
            color: #333;
        }

        body {
            max-width: 500px;
            margin: 0 auto;
            padding: 0 15px;
        }

        div,
        article {
            padding: 10px;
            margin: 5px;
            border: 1px solid #dedede;
        }
    </style>
</head>

<body>
    <h1>Accessing Elements in the DOM </h1>
    <h2>ID (#demo) </h2>
    <div id = "demo">Access me by ID </div>
    <h2>Class (.demo) </h2>
    <iv class = "demo">Access me by class (1)</div>
    <div class = "demo">Access me by class (2)</div>
    <h2>Tag (article) </h2>
    <article>Access me by tag (1)</article>
    <article>Access me by tag (2)</article>
    <h2>Query Selector </h2>
    <div id = "demo-query">Access me by query </div>
    <h2>Query Selector All </h2>
```

```
   <div class = "demo-query-all">Access me by query all
(1)</div>
   <div class = "demo-query-all">Access me by query all
(2)</div>
</body>

</html>
```

Now open this *anyName.html* file in the Chrome web browser, and open the browser's console. Now try the following commands and observe the changes:

```
//Finding element using ID
const demoId = document.getElementById('demo');
console.log(demoId);
demoId.style.border = '5px solid purple';
//observe purple border

//Finding element using Class Name

const demoClass = document.getElementsByClassName
('demo');
console.log(demoClass);
//observe collection of objects
demoClass[0].style.border = '5px solid orange';
demoClass[1].style.border = '5px solid red';

//Finding element using TAG
const demoTag = document.getElementsByTagName
('article');
console.log(demoTag);      //observe collection of objects
demoTag[0].style.border = '5px solid blue';
demoTag[1].style.border = '5px solid black';

//Finding element using querySelector (Single Element)
const demoQuery = document.querySelector('#demo-query');
//will only return first one to match
demoQuery.style.border = '5px solid orange';

//Finding element using querySelector (Collection)
const demoQueryAll = document.querySelectorAll('.demo-
query-all');
demoQueryAll.forEach(query =>{
   query.style.border = '1px solid green';
});           //using forEach and callback function
```

7.7 Event Handling Using DOM

We have now seen most of the properties and methods provided by DOM and how to use them to change the look and feel of the web page. The work done so far is not fruitful unless there is no active response mechanism within the website such as changing colors on the web page on some click, providing an interactive slider that computes on the go, hover over events, automatically play music on page loading and many more. Whatever we have learnt so far, we can do most of these manually, but now we will understand how to do all of this based upon some event such as a mouse click or a key press or the mouse rolling on the web page.

Events are some actions that occur in a browser either automatically or via user-initiated actions, such as on web page loading, click of a button, form submission, manual keypress, mouse click etc. The mechanism of handling these events is known as *event handling*. JavaScript provides *event handlers* and *event listeners* to assist and execute the event handling mechanism [102]. An *event handler* is a function that initiates when an event occurs. An *event listener* attaches a responsive interface to an element, which allows that particular element to wait and *listen* for the given event to fire.

The *document* object contains properties and methods related to event handlers such as animation, drag and drop, clipboard, full screen, load and unload, keyboard, pointer, selection, transition and touch. The list of event handlers provided by the language is very exhaustive, but the working principle is same for all of them. Let us first see some of the common HTML events, and then we will see few simple examples to observe them in action. By the end of this chapter, you will find a dedicated exercise for your hands-on event handling practice. These properties are part of the *document* object as every *Node* is derived from *EventTarget* class. These are the properties that can be attributed to elements along with a callback method that is called whenever the said event happens.

- *onchange*: This is triggered when the HTML element value is changed.
- *onclick*: This is triggered when the user clicks on the specified HTML element present inside the document.
- *onmouseover*: This event is triggered when the user moves the mouse over the specified HTML element.
- *onmouseout*: This event is triggered when the user moves the mouse away from the specified HTML element.
- *onkeydown*: This is triggered when the user pushes a keyboard key.
- *onload*: This event is triggered when the browser has finished loading the page.

Some examples:

- Open any web page, and in the console, type the following command. You will observe that clicking anywhere in the web page will give you an alert.

```
document.onclick = () => alert("hey")
```

- Try this example to create a button on a page, and clicking on this will print date inside the *<p>* element. Create *anyName.html* file and open it in Chrome browser.

```
<html>

<head>
   <title> Click Event Example </title>
</head>

<body>

   <button onclick = "document.getElementById('time').
innerHTML = Date()">The time is?</button>
   <p id = "time"> Click on button to see time here!!</p>
</body>

</html>
```

- Try this example to see how keyboard events are used. Create a new file using notepad, save it with anyName.html extension and open it using any web browser.

```
<html>

<head>
   <title> Click Event Example </title>
</head>

<body>
   <p id = "keyEvents"> Press any key on Keyboard to see
what you pressed!!</p>
</body>

</html>
<script>
```

```
document.addEventListener('keydown', event =>{
document.getElementById('keyEvents').innerHTML = 'You
pressed ' + event.key });
</script>
```

7.8 Exercise

7.8.1 Theory

    i. What is the difference between Node and Element interfaces used in DOM?

    ii. What do you mean by DOM? What role does DOM play while loading a web page in a web browser?

    iii. Explain the timing-based event handling mechanism in JavaScript with suitable demonstration using program code.

    iv. List out the different events used by DOM for event handling.

7.8.2 True/False

    i. Using * in a form field represents that the form field value cannot be left empty.

    ii. When a DOM tree is regenerated, the user is allowed to interact with the web browser.

    iii. *document* object is the parent for *window* object.

    iv. It is mandatory for *document* to have *form, anchor* and *paragraph* elements as its child nodes.

    v. Form elements in DOM can be accessed using the *form* object, which is a child of the *document* object.

7.8.3 Multiple-Choice Questions

    i. Everything is considered a(n) _____ in the Document Object Model.

       a. attribute

       b. node

       c. array

       d. string

ii. Objects within DOM are organized in a _____ manner.
    a. queue
    b. stack
    c. linked list
    d. hierarchical

iii. Which of the following is not used for event handling in JavaScript?
    a. onSubmit()
    b. onLoad()
    c. onFocus()
    d. hover()

iv. In *document* object, _____ are included.
    a. form
    b. *<p>* elements
    c. links
    d. All of the above

v. Chose the correct statement from the following:
    a. DOM follows the W3C standard.
    b. A single web page can have multiple *document* objects.
    c. Node and Elements can be used interchangeably in DOM.
    d. All of the above

7.9 Demo and Hands-On DOM

7.9.1 Objective

i. Understand DOM, and understand its use via code.

ii. Execute basic operations on DOM.

iii. React to events on a web page via JavaScript.

7.9.2 Prerequisite

- **Step 1:** Visit https://www.routledge.com/9780367641429 and download zip file for *Ch7_DOM_Template.*

- **Step 2:** Unzip the content and open *index.html* in any editor of your choice (example: Notepad or Visual Studio Code).
- **Step 3:** Insert code given next inside <**script**> block in *index.html*.
- **Step 4:** Repeat the above three steps for each snippet given next.

7.9.3 Explore

Using any editor, we will write and test JavaScript code to interact with the DOM and see the results on the web page. Here's what you should know about DOM. DOM stands for Document Object Model. It is a part of the browser window and is the bridge through which JavaScript interacts with HTML and CSS.

7.9.3.1 Code Snippet-1

```
/* Creating a reference to any element of HTML document.
 * Syntax
 * const REFERENCE_NAME = document.querySelector
(<.class/#id>)
 */

// For instance, creating a reference to id 'p1'
const myP1 = document.querySelector("#p1 p");
// Reference to id 'p2'
const myP2 = document.querySelector("#p2 p");

// Try printing out into console, to see what all it holds
console.log("P1", myP1);
console.log("P2", myP2);
// Also, lets create and array of messages (It will come
handy soon.)
let myMessages = [
    "Hello there! Coder.",
    "Hope you having an amazing day <3",
    "Have patience",
    "Rome wasn't built in a day:)",
];

// Manipulating the content inside HTML element,
// * Method-1 | Changing the existing content.
myMessages.forEach((msg) => {
    // for loop iterates through every element
    myP1.innerHTML = msg; // of myMessages, msg is the loop
variable.
```

```
});

/******************** DISPLAYS INFO ****************
******** */
let contentInfo = document.createElement("p");
contentInfo.innerText = "I'm overWritten content!";
contentInfo.style.backgroundColor = "pink";
myP1.appendChild(contentInfo);

// Notice only last message is visible in p1 content,
// That's because it's overwritten in every loop iteration.
// That's not what we want, so here's…

// * Method-2 | Appending to existing content.
myMessages.forEach((msg) => {
    // Since we want to append some text, let's create a text
element.
    // * Syntax
    // * newElement = document.createElement(<element-tag>)
    // * newElement.innerText = "some text", assign some text
value to new element.
    let myContent = document.createElement("p");
    myContent.innerText = msg;
    myP2.appendChild(myContent);
});

/******************** DISPLAYS INFO ****************
****** */
contentInfo = document.createElement("p");
contentInfo.innerText = "I'm appended content!";
contentInfo.style.backgroundColor = "limegreen";
myP2.appendChild(contentInfo);
```

7.9.3.2 Code Snippet-2

```
/* Creating a reference to any element of HTML document.
 * Syntax
 * const REFERENCE_NAME = document.querySelector
(<.class/#id>)
 */

// For instance, creating a reference to <p>inside id 'p1'
const myP1 = document.querySelector("#p1 p");
```

```javascript
// Same for id 'p2'
const myP2 = document.querySelector("#p2 p");

// Reference to button | class 'btn' inside id '#p1'.
const p1Btn = document.querySelector("#p1.btn");
// Same for #p2
const p2Btn = document.querySelector("#p2.btn");

// Try printing out into console, to see what all it holds
console.log("P1", myP1);
console.log("P2", myP2);

// Also, lets create and array of messages (It will come
handy soon.)
let myMessages = [
    "Hello there! Coder.",
    "Hope you having an amazing day <3",
    "Have patience",
    "Rome wasn't built in a day:)",
    "This session is powered by electromagnetic-cheese!",
];

p1Btn.addEventListener("click", () => {
    myMessages.forEach((msg) => {
        // for loop iterates through every element of
myMessages, msg is the loop variable.
        myP1.innerHTML = msg;
    });
    // Look into the console, to understand the change in HTML
element.
    console.log("P1", myP1);
});

// Instead of defining your reaction-callback() within event-
listener param. list,
// you can pass the reference to the function too, for
instance…
const appendContent = () => {
    myMessages.forEach((msg) => {
        // Since we want to append some text, let's create a
text element.
        // * Syntax
        // * newElement = document.createElement
(<element-tag>)
        let myContent = document.createElement("p");
```

```
    // * newElement.innerText = "some text", assign some
text value to new element.
    myContent.innerText = msg;

    // Append to HTML reference
    myP2.appendChild(myContent);
});
```

```
// Look into the console, to understand the change in HTML
element.
    console.log("P2", myP2);
};
```

```
// Let's add event listener to button in #p2
p2Btn.addEventListener("click", appendContent);
```

7.9.3.3 Code Snippet-3

```
/* Creating a reference to any element of HTML document.
 * Syntax
 * const REFERENCE_NAME = document.querySelector
(<.class/#id>)
 */
```

```
// For instance, creating a reference to id 'p1'
const myP1 = document.querySelector("#p1");
// Reference to id 'p2'
const myP2 = document.querySelector("#p2");
```

```
// Reference to button, inside class 'results' | class 'btn'.
const myButton = document.querySelector(".results.btn");
// Reference to p3, (To display results)
const resultField = document.querySelector
(".results #p3");
```

```
// Try printing out into console, to see what all it holds
console.log("P1", myP1);
console.log("P2", myP2);
console.log("Result-Field", resultField);
```

```
// Also, lets create and array of messages (It will come
handy soon.)
let myMessages = [
    "Hello there! Coder.",
    "Hope you having an amazing day <3",
```

```
      "Have patience",
      "Rome wasn't built in a day:)",
      "This session is powered by electromagnetic-cheese!",
];

let myColors = [
      "limegreen",
      "red",
      "blue",
      "black",
      "tomato",
      "crimson",
      "majenta",
];

// Function to return a random element from array
const getRandomElement = (arr) => {
      return arr[Math.floor(Math.random() * arr.length)];
};

// Let's add an event listener to myButton
// * Syntax
// <variable-name>.addEventListener("<event-that-you-
want-to-track>", () => {…def your function for reaction
to event… })
myButton.addEventListener("click", () => {
      // Use DOM to change background color of div-class
'results'
      document.querySelector
(".results").style.backgroundColor = getRandomElement(
         myColors
      );

      // Assign a random message into result field.
      resultField.innerHTML = getRandomElement(myMessages);
});
```

7.9.3.4 Code Snippet-4

```
/* Creating a reference to any element of HTML document.
 * Syntax
 * const REFERENCE_NAME = document.querySelector
(<.class/#id>)
 */
```

```
// For instance, creating a reference to <p>inside id 'p1'
const myP1 = document.querySelector("#p1");
// Same for id p2, p3, p4, p5;
const myP2 = document.querySelector("#p2");
const myP3 = document.querySelector("#p3");
const myP4 = document.querySelector("#p4");
const myP5 = document.querySelector("#p5");
// Utilities
const ready = document.getElementById("prepare");
const btnRow = document.querySelector(".button-row");
const render = document.querySelector(".factory-render");
const setSrc = document.getElementById("setSrc");
const getSrc = document.getElementById("getSrc");
const img = document.getElementById("image");
const pop = document.querySelector(".pop");
const scrollRender = document.querySelector("#p5.factory-
render");

/* INTERACTING WITH HTML CLASS ATTRIBUTES ADDING/REMOVING
 * HINT: A css property for class 'hide' has been defined al-
ready in style.css
 * It sets the display to none.
 */

// Adding a class to a DOM element.
// p1 and p2 were used for previous snippets, let's try to
hide them.
// and also, we will be needing p3, p4, p5 for this experiment
(Unhide them).
// Lets use a button to do this setup for us.

// Unhiding the button.
ready.classList.remove("hide");
ready.addEventListener("click", (e) => {
    // HIDE p1,p2 and the button itself.
    myP1.classList.add("hide");
    myP2.classList.add("hide");
    ready.classList.add("hide");
    // Unhide p3, p4, and p5.
    myP3.classList.remove("hide");
    myP4.classList.remove("hide");
    myP5.classList.remove("hide");
});
```

```javascript
// Let's fuel the HTML factory.
btnRow.addEventListener("click", (e) => {
    if (e.target.tagName = == "BUTTON") {
    console.log("Click registered.");

    // Pick the first name in the list of class.
    const clicked = e.target.classList[0];

    // Let's see class button belonged to -->
    if (clicked = == "p") {
        //
        // Paragraph creation requested.
        // Will use createElement() method to create one.
        var newP = document.createElement("p");
        newP.innerText = "I'm random text, have no sense or
reason to exist."; // Add some text content to new <p>element.
        // Add some style
        newP.style.color = "#34656d";
        newP.style.backgroundColor = "#c6ffc1";
        // It's ready, but needs to be pushed into the HTML doc.
        render.appendChild(newP);
        //
        //
    } else if (clicked = == "d") {
        //
        // Creating a Data input Element.
        var newD = document.createElement("input");
        newD.type = "date";
        render.appendChild(newD); // Append
        //
        //
    } else if (clicked = == "a") {
        //
        // Creating an anchor tag.
        var newA = document.createElement("a");
        newA.href = "https://youtube.com";
        newA.innerText = "YouTube";
        newA.style.color = "white";
        newA.style.backgroundColor = "red";
        newA.style.borderRadius = "15%";
        render.appendChild(newA); // Append
        //
        //
    } else if (clicked = == "c") {
        //
```

```
        // Creating a new circle block.
        var newC = document.createElement("div");
        newC.style.width = "50px";
        newC.style.height = "50px";
        newC.style.backgroundColor = "seagreen";
        newC.style.borderRadius = "50%";
        render.appendChild(newC);
    } else {
    return;
    }
}
});


/********************** GET / SET attributes. **********
**********/
// This URL id guaranteed to return a random square
// sized-image of 200-pixels wide.
const randomImageURL = "https://picsum.photos/200";
// const randomImageURL = "https://picsum.photos/400/200";
// For rectangle-sized image, feed any dimensions

setSrc.addEventListener("click", (e) => {
    // Set 'src' attribute to a valid image source.
    img.setAttribute("src", randomImageURL);
});

getSrc.addEventListener("click", (e) => {
    document.querySelector(".get").innerText = img.
getAttribute("src");
});


/********************** Scrolling with javascript******
******************/
pop.addEventListener("click", (e) => {
    for (var i = 0; i <1000; i + +) {
        var newP = document.createElement("p");
        newP.innerText = "I'm random text, have no sense or
reason to exist."; // Add some text content to new <p>element.
        // Add some style
        newP.style.color = "#34656d";
        newP.style.backgroundColor = "#c6ffc1";
        // It's ready, but needs to be pushed into the HTML doc.
        scrollRender.appendChild(newP);
    }
});
```

```
document.getElementById("scrollDown").addEventListener
("click", (e) => {
    // window.scrollTo(x,y);
    // This methods let's the current view of the screen to be
shifted
    // to any (x,y) co-ordinates.
    // PROTIP: document.body.scrollHeight returns the bottom
co-ordinate.

    window.scrollTo(0, document.body.scrollHeight);
});

document.getElementById("scrollUp").addEventListener
("click", (e) => {
    // window.scrollTo(), also takes an object, wherein
options, and
    // relative co-ordinates like top, left, right, bottom,…
can be
    // provided.
    window.scrollTo({
        top: 0,
        left: 0,
        behavior: "smooth",
    });
});
```

7.10 Demo and Hands-On for Event Delegation

7.10.1 Objective

i. Understand the alternative workaround instead of event delegation and how painful it is.

ii. Event Delegation example.

7.10.2 Prerequisite

- **Step 1:** Visit https://www.routledge.com/9780367641429 and download zip file for *Ch7_EventDelegation_Template*.
- **Step 2:** Unzip the content and open *index.html* in any editor of your choice (example: Notepad or Visual Studio Code).

- **Step 3:** Insert code given next inside <**script**> block in *index.html*.
- **Step 4:** Repeat the above three steps for each snippet given next.

7.10.3 Explore

Using any editor, we will write and test JavaScript code to interact with the DOM and see the results on the web page. Here's what you should know about Event Delegation.

Capturing and bubbling allow us to implement one of the most powerful event handling patterns called *event delegation*.

The idea is that if we have a lot of elements handled in a similar way (listening to similar events); then instead of assigning a handler to each of them, we put a single handler on their common ancestor (parent-container).

In the handler we get *event.target* to see where the event actually happened and handle it.

7.10.3.1 Code Snippet-1

```
// Utility ()
const appendToHTML = (contents, color = "white") => {
contents.forEach((content) => {
  const myPara = document.createElement("p");
   myPara.style.color = color;
   myPara.innerHTML = content;
   resultHere.appendChild(myPara);
   sep();
  });
};

const sep = () => {
  const separator = document.createElement("hr");
  resultHere.appendChild(separator);
};
/*****INEFFICIENT WAY TO DO | TRY THIS OUT AND CHECK FOR YOURS-
ELF THE PAIN IN THIS METHOD <3*****************************
***********************/

// Let's render some colors to the boxes
const colors = ["red", "green", "blue", "magenta", "cyan"];
var clr = 0;
const ancestorElement = document.querySelector(".blobs");
var childNodes = ancestorElement.childNodes;
for (let i = 0; i <childNodes.length; i + +) {
   if (childNodes[i].nodeName.toLowerCase() === "button") {
```

```
    childNodes[i].style.background = colors[clr + +];
  }
}

// Goal is to check, which color was clicked.
var resultHere = document.querySelector(".results");
appendToHTML([
  "Goal is to check, which color was clicked",
  "Adding separate event listeners to all 5 buttons",
]);

appendToHTML(
  [
    "'event' parameter is passed into the callback function
of event-listener",
    "'target' property of 'event' gives the exact object on
which the event happened!",
    "using 'classList' property of 'target' we extract the
name of the class of the target.",
  ],
  "#ff574c"
);
// Red Listener
document.querySelector(".red").addEventListener
("click", (event) => {
  alert(`You clicked ${event.target.classList} <3`);
});
// Green Listener
document.querySelector(".green").addEventListener
("click", (event) => {
  alert(`You clicked ${event.target.classList[0]} <3`);
});
// Purple Listener
document.querySelector(".blue").addEventListener
("click", (event) => {
  alert(`You clicked ${event.target.classList[0]} <3`);
});
// Blue Listener
document.querySelector(".purple").addEventListener
("click", (event) => {
  alert(`You clicked ${event.target.classList[0]} <3`);
});
// Cyan Listener
document.querySelector(".cyan").addEventListener
("click", (event) => {
```

```
    alert(`You clicked ${event.target.classList[0]} <3`);
});
```

7.10.3.2 Code Snippet-2

```
// Utility ()
const appendToHTML = (contents, color = "white") => {
    contents.forEach((content) => {
        const myPara = document.createElement("p");
        myPara.style.color = color;
        myPara.innerHTML = content;
        resultHere.appendChild(myPara);
        sep();
    });
};

const sep = () => {
    const separator = document.createElement("hr");
    resultHere.appendChild(separator);
};
/***** <EFFICIENT WAY TO HANDLE SIMILAR EVENTS ON MULTIPLE
EVENTS>********************************************
*******/

// Let's render some colors to the boxes
const colors = ["red", "green", "blue", "magenta", "cyan"];
var clr = 0;
const ancestorElement = document.querySelector(".blobs");
var childNodes = ancestorElement.childNodes;
for (let i = 0; i <childNodes.length; i + +) {
    if (childNodes[i].nodeName.toLowerCase() === "button") {
        childNodes[i].style.background = colors[clr + +];
    }
}

// Goal is to check, which color was clicked.
var resultHere = document.querySelector(".results");
appendToHTML([
    "Goal is to check, which color was clicked",
    "Adding single event listener to the ancestor (parent)
container <3",]);

sep();
appendToHTML(
    [
```

```
        "Since, no buttons have individual event handlers, the
event will BUBBLE up and CAPTURED by the event-handler of
parent, this is EVENT-DELEGATION <3",
    ],
    "black"
);
sep();

appendToHTML(
    [
        "'event' parameter is passed into the callback func-
tion of event-listener",
        "'target' property of 'event' gives the exact object
on which the event happened!",
        "using 'classList' propert of 'target' we extract the
name of the class of the target.",
        "This event-handler will now respond to any click,
even on text or blank surface, let's use conditional to check
for that.",
    ],
    "#ff574c"
);

// Red Listener
document.querySelector(".blobs").addEventListener
("click", (event) => {
    // This condition makes sure, only button clicks are re-
sponded, nothing else. Comment this out to verify the other
scenario.
    if (event.target.nodeName.toLowerCase() === "button") {
        alert(`You clicked ${event.target.classList} <3`);
    } else {
        return;
    }
});
```

8

Standard Built-In Objects

Collecting is more than just buying objects.

—Eli Broad

In Chapter 3, we have learned about the internals of object-oriented programming (OOP) in JavaScript including the creation of objects, deletion of objects, prototypal inheritance and garbage collection. We have also learned about the concept of a class that helps to make as well as extend from predefined entities. While explaining the fundamentals of objects and their extension, we felt the need of explaining the built-in objects that are supported by JavaScript. These objects can be considered as the most commonly used choice by developers used for extending functionality of newly created objects. This will help you get more control as well as power over your web page than any other application for that matter. In this chapter, we discuss some of the standard built-in objects provided by ECMAScript6 [103] in a little bit more depth.

8.1 Built-In Objects

JavaScript offers many standard objects that aim at assisting the developer to gain more control over the web page and allow implementation of more user-interactive solutions. These objects are built in the language and allow inheritance on them, thereby extending functionalities to objects further inheriting from them. Standard built-in objects are also sometimes referred to as *global objects* because of their visibility across the global scope of the script.

Apart from built-in objects provided by JavaScript, other global access objects are also available such as the *window* object in BOM and the *document* object in DOM. Built-in JavaScript objects are not related to BOM and DOM object models but are available to be used together to build a more cohesive web page or application. The built-in objects are classified as shown in Figure 8.1.

DOI: 10.1201/9781003122364-8

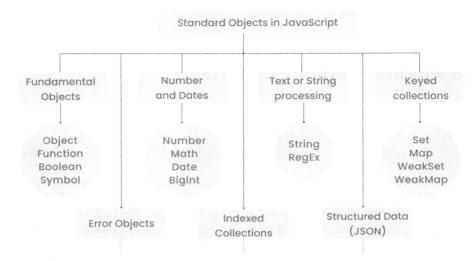

FIGURE 8.1
Classification of standard built-in JavaScript objects.

8.1.1 Fundamental Objects

Fundamental objects are the parent to most of the user-created objects. Mostly all the user-created objects inherit from these fundamental objects. The fundamental objects include *Object, Function, Boolean* and *Symbol* classes. Out of these fundamental objects, we have already discussed Object and Function at great length in the previous chapters. These fundamental objects are briefly discussed in the following section. We are intentionally limiting the discussion on Object and Function here as they are covered separately in the previous sections.

8.1.1.1 Object Prototype

Object prototype is the root of prototypal chain in JavaScript. Many other object prototypes inherit from the generic Object prototype, including Function object. Whenever the *new* keyword is used to create an object, it automatically inherits from *Object prototype* and thus inherits the properties as well as methods from this prototype [104]. These properties and methods provide added functionality to an object as compared with an object that is created using the Object literals.

Prototype is a global object constructor that is used to add properties (including methods) to any object inheriting from *Object Prototype*. Object prototype has some built-in methods such as *toString()* method, which returns a string holding the value of object. This method can be overridden when needed by objects that are inheriting from *Object()*.

```
var obj1 = new Object();
//inherits from Object prototype
var obj2 = [name:"Syngonium", color:"light green"];
//does not inherit
obj2.toString();
//gives error as the method is not available for obj2
obj1.toString();
//works fine
```

8.1.1.2 Function Object

By default, every user-defined function includes a Function prototype object. Every function in JavaScript inherits from the Function object that provides a *constructor, instance properties* and *methods* [105]. Adding user-defined methods to Function prototype ensures that all objects also get a copy of the same function down the prototype chain. This adds more power to the methods being defined via the Function prototype.

Using the *Function()* constructor creates a new Function object that works within the global scope. Instance properties for Function Object are *Function.prototype.name()* that returns the name of the function, *Function.prototype.displayName* that returns the name of the function and *Function.prototype.length* that returns the number of arguments belonging to the function.

Instance methods belonging to Function object include *toString(), apply(), bind()* and *call(). Function.prototype.toString()* returns the string representation of the function's source code. It overrides the method provided by the Object prototype. *Function.prototype.apply(newThis)* replaces the calling object's *this* argument with the *newThis* argument. *bind()* and *call()* method are used to create a new function and replaces its *this* argument with the passed argument. *call()* is used to call a function and then set its *this* argument. As function finds a very critical position in JavaScript development, these concepts were already discussed before arriving here.

8.1.1.3 Boolean Object

Traditional boolean values hold either *true* or *false* as their value, i.e. they are primitive in nature. A *Boolean object* belongs to the Object prototype and serves as an object wrapper for holding a boolean value. The value of a Boolean object is implicitly set to *false* for values 0, −0, null, false, NaN and undefined, and it is set to *true* for all other values [106]. The following example demonstrates numerous ways to create a Boolean object with *false* and *true* values, respectively.

```
let obj1 = new Boolean();      //value is false
let obj2 = new Boolean(0);     //value is false
```

```
let obj3 = new Boolean(null);
//null also sets the value as false
let obj4 = new Boolean(' ');           //resolves to false
let obj5 = new Boolean(false);    //resolves to false
let obj6 = new Boolean(true);     //resolves to true
let obj7 = new Boolean('true');      //true
let obj8 = new Boolean('false');
// true, treated as a string
let obj9 = new Boolean('T&F');
// true, treated as a string
let obj10 = new Boolean([]);      //true
let obj11 = new Boolean({});      //true
```

Using Boolean object in decision and control flow statements adds some confusion; you will notice this while practicing with boolean primitives and Boolean object. Any Boolean object holding a value other than *undefined*, *null* or a Boolean object with value set to *false*; will return *true* when passed to any type of conditional statement. Read that again. Let us understand this peculiar concept in action using the following example:

```
let b = new Boolean(false);
    //assigning Boolean object to 'b' reference

let c = false;               //Boolean value as primitive
if (b)                  //evaluates it as true
{
    //the code gets executed as b resolves to true
}
if (c)                  // evaluates it as false
{
    //code is not executed as c resolves to false
}
```

Therefore, one must use boolean primitives and Boolean object carefully. As shown in the previous example, the initial value of a Boolean object resolves to *true* even if it is set as *false*.

The Boolean prototype provides a constructor that allows adding properties or methods to the prototype itself. Any change to the prototype will be inherited by all the objects down the chain. Syntax for using *Boolean()* prototype constructor is given as follows:

```
Syntax:
Boolean.prototype.name = value
```

Boolean.prototype refers to the *Boolean()* object itself. In the above syntax declaration, *name* refers to the key and *value* refers to the actual value of the property to be set. The same notation is used for adding methods to *Boolean()*. Boolean prototype comes with two built-in methods: *toString()* and *valueOf()*. *toString()* method returns either *true* or *false* depending upon the calling object. It overrides the method provided by Object prototype. In a similar manner, objects inheriting from *Boolean()* can also override this method and provide a different implementation. Another method originally belonging to *Object()* and overridden by *Boolean()* is the *valueOf()* method, which returns the primitive value of *true* or *false* depending upon the calling object.

8.1.1.4 Symbol Object

Symbol object is another built-in fundamental object for JavaScript, which aims at uniquely identifying properties for its objects. It helps in achieving a weak encapsulation by providing values, which other code-executing mechanisms would not use [107]. Constructor for Symbol Object always returns a *symbol* primitive that is unique across all other properties.

Symbol object comes with its own constructor to create a new *Symbol* object. However, use of *new* keyword is not supported by the Symbol prototype. Let us understand with help of an example:

```
let symbol_A = Symbol();       //creates a new Symbol object
let symbol_B = Symbol('JavaScript');
// optional string value
let symbol_C = Symbol('JavaScript');
//another new object
let symbol_X = new Symbol();
//throws TypeError because of new
console.log(symbol_B = = = symbol_C)    //prints false
typeof Symbol() === 'symbol';           //returns true
typeof Symbol('JavaScript') === 'symbol' //returns true
typeof symbol_B = = = 'symbol';         //returns true
```

In the above example, three objects of Symbol are created, *symbol_A*, *symbol_B* and *symbol_C*. The fourth object *symbol_X* throws Type Error because Symbol does not support the usage of *new* operator. Symbol ensures that new object creation happens via *Object()* constructor so that the uniqueness is maintained across these objects, which is their key defining purpose for existence. Further, observe that the statement *symbol_B = = = symbol_C*; returns *false* even when both the objects are created using the same string. *typeof* operator can be used to identify the nature of these objects, i.e. whether they are Symbol objects or not, as demonstrated in the example.

Object.getOwnPropertySymbols() method is used to return an array of Symbol properties being defined on the calling object. Symbol prototype is not inherited on its own unless explicitly executed by the user. Therefore, a user-defined object (not inheriting from Symbol) will not have any Symbol properties unless they are set by the program.

Symbol object comprises certain in-built instance properties and methods as well. Instance property includes *Symbol.prototype.description* that returns the description of calling Symbol object in the form of a read-only string. Instance methods include *toSource(), toString()* and *valueOf()* that are overridden implementations of the *Object* prototype's method. *Symbol.prototype.toString()* is used to return the string value of Symbol object's description. *Symbol.prototype. valueOf()* is used to return the Symbol, whereas *Symbol.prototype.toSource()* is used to return a string holding the source of the calling Symbol.

8.1.2 Error Objects

The JavaScript runtime throws a special object whenever any runtime error occurs; this special object is called an *Error* object [108]. The *Error* object is a fundamental object that includes basic as well as special error types such as *EvalError, RangeError, ReferenceError, AggregateError, SyntaxError, TypeError and URIError*. The error objects are used for exception handling and are discussed in detail in Chapter 8. The error object has two properties only, i.e. *Error.prototype.message* and *Error.prototype.name,* both of which are used to store the message about error and name of that error, respectively. The Error object only contains one method, i.e. *Error.prototype.toString()* method, which is an overridden implementation inherited from the Object class. The error object is used to provide exceptional handling in JavaScript by using keywords like *try, catch, throw* and *finally*.

8.1.3 Number and Dates

Number and dates are used to work around numbers, dates and other mathematics-based calculations. These include *Number, Math, Date and BigInt* objects. These objects are really helpful while working with web content, which needs to work globally across different time zones.

8.1.3.1 Number Object

Number object contains various constants and methods used to work with numbers [109]. Number value in JavaScript is equivalent to double precision 64-bit binary format, just like *double* in Java. It can keep precision up to 17 decimal places, post which the number is considered as equivalent to the constant *Infinity*.

Number object has the *Number()* constructor, which is used to generate a new *Number* value. *Number(value)* as a method returns the numeric value corresponding to the string or any other value being passed. If the *value* is not convertible to Number, then it returns *NaN*. For example:

```
let year = 2021;
let previousYear = Number('2021');
y = == z;                          //returns True
Number('twenty twenty - one')      //returns NaN
```

The Numbers object have many static properties, static methods, instance properties and instance methods. Some of the properties belonging to this object are explained next:

- *Static properties: Number.NaN* represents and returns the value *Not a Number. Number.EPSILON* returns the difference or the smallest interval existing between 1 and the given number. It is often used to test the equality between two floating point numbers. *Number. MAX_SAFE_INTEGER* and *Number.MIN_SAFE_INTEGER* represent the value of maximum ($2^{53} - 1$) and minimum safe integer ($-(2^{53} - 1)$), respectively. *Number.MAX_VALUE* and *Number.MIN_VALUE* represent the largest positive number and the smallest positive number (closest to zero), respectively. *Number.NEGATIVE_INFINITY* and *Number.POSITIVE_INFINITY* return a special value representing negative infinity and positive infinity, respectively. Often, they are used for dealing with overflow conditions. *Number.prototype* property is used to add more properties to the *Number* object itself.

- *Static Methods: Number.isNan()* returns true if the passed number is NaN; else it returns false. This method is different from global *isNaN(). isNan()* converts the passed value to a number and then tests its equivalence to NaN, whereas *Number.isNaN()* does not convert the passed value to a number and simply determines its equivalence to NaN. *Number.isFinite()* returns true if the passed number is finite in nature; else it returns false. *Number.isInteger()* returns true if the passed value is an integer; else it returns false.

- *Instance Method: Number.prototype.toExponential(digitFraction)* is used to convert the number into exponential form and returns a string value for it. *Number.prototype.toFixed(digits)* converts the calling number to a string by rounding off the decimals as per the *digit* argument. If the value of *digits* or decimal places is higher than the calling number, zeroes are appended to the string in order to create the desired length. *Number.prototype.toPrecision(value)* returns the string representation of the number to specified precision value. For example:

```
let x = 1.12891;
let y = x.toFixed(3);    //rounded off to 3 decimal places
console.log(y);          //prints 1.129 to console
let z = x.toFixed(8);    //rounds off to 8 decimal places
console.log(z);          //prints 1.12123000
```

Number.prototype.toString(base) returns the string representation of the calling object as per the specified base (or radix value). *Number.prototype.valueOf()* returns the primitive value held by the calling object. Both these methods are overridden implementations of *Object.prototype* methods.

8.1.3.2 Date Object

Date Object is used to represent a particular moment of time in a system-independent way. Date Object inherently uses a Number that represents total number of milliseconds passed/elapsed since January 1, 1970 universal time coordinated (UTC), also called UNIX epoch [110]. This helps in achieving easier date and time zone conversions across multiple geographic zones. Date object offers methods to convert and display time in standard formats such as UTC and Greenwich mean time (GMT). A user's platform, i.e. a browser in this case, provides the local time that is then converted and stored by Date Object [111].

Date Object offers two types of constructors, i.e. *Date()* and *new Date()*. Creating an object via *new Date()* returns a new Date object, whereas simply calling *Date()* as a function returns the current string representation of date and time belonging to the user's system [112].

```
let currentDate = new Date();        //creates a new object
let travelDay = new Date('March 29, 2021 12: 11: 10');
//creating new object with a specified date
let travelDay = new Date(2021, 3, 29)
//creates a new object with same date
let travelDay = new Date(2021, 3, 29, 12, 11, 10)
//another way to pass arguments
```

Static methods belonging to Date class include *Date.now()*, *Date.parse()* and *Date.UTC()*. *Date.now()* returns a numeric value representing the number of total milliseconds passed since January 1, 1920 (UTC). Leap seconds are ignored by this method. *Date.parse()* returns a string representation of date along with the number of milliseconds elapsed since standard base time. *Date.UTC()* returns the difference between the date passed as an argument and the standard base time.

The following list outlines some of the commonly used instance methods belonging to *Date.prototype*.

- *getDate() and getUTCDate():* These methods return the day of the month (range: 1–31) as per the local user time and the universal time, respectively.

- *getDay() and getUTCDay():* These methods return the day of the week (range: 0–6) as per local user time and the universal time, respectively.

- *getFullYear() and getUTCFullYear():* These methods return a four-digit number representing the year as per local time and universal time, respectively.

- *getHours() and getUTCHours():* These methods return the current hour (range: 0–23) as per local user time and universal time, respectively.

- *getMilliseconds() and getUTCMilliseconds():* These methods return the number of milliseconds elapsed in range 0–999 as per local user time and universal time, respectively.

- *getMinutes() and getUTCMinutes():* These methods return the number of minutes in range 0–59 as per the local user time and universal time, respectively.

- *getMonth() and getUTCMonth():* These methods return the number corresponding to the current month (range 0–11) as per the local user time and universal time, respectively.

- *getSeconds() and getUTCSeconds():* These methods return the number of seconds elapsed on specified date (range 0–59) as per the local user time and universal time, respectively.

- *getTime():* This method returns a numeric value corresponding to the total number of milliseconds passed since the standard time, i.e. January 1, 1970 UTC.

- *getTimezoneOffset():* This method returns the offset for time zone belonging to the user's local time.

- *setDate() and setUTCDate():* These methods are used to set the day of the month as specified for the user's local time and universal time, respectively.

- *setFullYear() and setUTCFullYear():* These methods are used to set the year (four-digit number) as per the user's local time and universal time, respectively.

- *setHours() and setUTCHours():* These methods are used to set the hour (range: 0–23) for the specified date as per the user's local time.

- *setMilliseconds() and setUTCMilliseconds():* These methods are used to set the milliseconds for the specified date as per the user's local time and universal time, respectively.

- *setMinutes() and setUTCMinutes():* These methods are used to set the minutes (range: 0–59) for the specified date as per the user's local time and universal time, respectively.

- *setMonth() and setUTCMonth():* These methods are used to set the month for the specified date as per the user's local time and universal time, respectively.

- *setSeconds() and setUTCSeconds():* These methods are used to set the number of seconds for the specified date as per the user's local time.

- *setTime():* This method is used to set the Date object to the specified value.
- *toDateString():* This method returns the date value from Date object in a string format (human readable form).
- *toJSON():* This method returns a string value representing the Date object using toISOString() method.
- *toLocaleDateString():* This method returns the string value for date corresponding to the user's local system settings.
- *toLocaleString():* This method overrides the Object's *toLocaleString()* method to return the string comprising date as per the local system settings.
- *toLocaleTimeString():* This method returns a string representing the local time depending on the user's system settings.
- *toString():* This method returns the string value of the calling Date object. It overrides the Object prototype's *toString()* method.
- *valueOf():* This method returns the primitive numeric value held by the calling Date object. It overrides the Object prototype's *valueOf()* method.

By using the aforementioned methods, Date object can be easily used to set, retrieve and manipulate these values as per the need of the program. Date object is more commonly used with the Number object as demonstrated using the example given next:

```
let d = new Date('April 17, 2021 01: 22: 44')
console.log(Number(d));       //converting Date to Number
```

Date class provides many instance methods that can be easily used to manipulate and work around date- and time-related operations. These methods are heavily used to create a unified time- and date-sensitive web content for global users.

8.1.3.3 Math Object

Math object is another built-in object that provides properties and methods needed for complex mathematical operations and constants [113]. You cannot create an object of Math as Math Object does not provide any constructor; rather it provides static methods and properties that can be globally used with ease. Most of these are constants that are used to supply the given values to the program. These properties have a platform or system-dependent implementation; therefore, different browsers can give different results while executing methods from Math Object. The following list shows most of the static properties available in Math object followed by the static methods:

- *Math.E*: This static property returns the numeric value of Euler's constant and the base of natural logarithms, i.e. 2.718.
- *Math.LN2*: This static property returns natural logarithmic value of 2, i.e. 0.693.
- *Math.LOG2E*: This static property returns Base-2 logarithmic value of E, i.e. 1.443.
- *Math.LOG10E*: This static property returns Base-10 logarithmic value of E, i.e. 0.434.
- *Math.PI*: This static property returns mathematical PI constant, i.e. 3.14159.
- *Math.SQRT1_2*: This static property returns square root of ½, i.e. 0.707.
- *Math.SQRT2*: This static property returns square root of 2, i.e. 1.414.

Static Methods available inside Math object are as follows:

- *Math.abs(X)*: This static method returns absolute value of X.
- *Math.acos(X)*: This static method returns the arc-cosine value for X.
- *Math.acosh(X)*: This static method returns the hyperbolic arc-cosine for X.
- *Math.asin(X)*: This static method returns the arcsine value for X.
- *Math.asinh(X)*: This static method returns the hyperbolic arcsine for X.
- *Math.atan(X)*: This static method returns the arc-tangent value for X.
- *Math.atanh(X)*: This static method returns the hyperbolic arc-tangent value for X.
- *Math.cbrt(X)*: This static method returns the cube root value for X.
- *Math.ceil(X)*: This static method returns the value of smallest integer equal to or greater than X.
- *Math.floor(X)*: This static method returns the value of largest integer equal to or less than X.
- *Math.cos(X)*: This static method returns the cosine value for X.
- *Math.cosh(X)*: This static method returns the hyperbolic cosine value for X.
- *Math.sin(X)*: This static method returns the sine value for X.
- *Math.sinh(X)*: This static method returns the hyperbolic sine value for X.

- *Math.tan(X):* This static method returns the tangent value for X.
- *Math.tanh(X):* This static method returns the hyperbolic tangent value for X.
- *Math.exp(X):* This static method returns the value of E^X where E is Euler's constant, i.e. 2.718.
- *Math.log(X):* This static method returns the value of natural logarithm for X.
- *Math.log2(X):* This static method returns the value of base-2 logarithm for X.
- *Math.log10(X):* This static method returns the value of base-10 logarithm for X.
- *Math.max(a,b,c,d...) and Math.min(a,b,c...):* These static methods return the largest and the smallest number, respectively, from the list of arguments passed.
- *Math.pow(X,Y):* This static method returns the value of base X raised to the power Y, i.e. X^Y.
- *Math.random():* This static method returns a random real number between 0 and 1.
- *Math.round(X):* This static method rounds off the value of X to nearest integer and returns its value.
- *Math.sign(X):* This static method returns the sign value (i.e. positive, negative or zero) for X.
- *Math.trunc(X):* This static method returns the integer part of X by removing fractional part belonging to X.

8.1.4 Text or String Processing

Text or String processing is achieved in JavaScript by using either the *String* object or the *RegExp* object. String is also primitive in JavaScript. Let us understand them in detail in this section.

8.1.4.1 *String Object*

String Object is used to handle, represent and manipulate a series of characters that hold data in textual format. Just like other primitive objects, JavaScript distinguishes between primitive string values and the String objects [114]. When String Object methods are called on primitive strings, JavaScript automatically do type conversion and then calls the specified method.

Strings can be created via primitive values, via string literals or via *String()* constructor belonging to String Object. *String()* constructor is used to create a new String object, or it can be called a function to conduct automatic type conversion. Individual characters can be accessed from the String Object by using *charAt()* method or via array-like indexing as demonstrated in the next example:

```
let str1 = 'string value';    //creating new String object
let str2 = new String("A new object");
//using String() constructor
let str3 = str1.charAt(3);
//str3 now holds 'i' as its value
let str4 = str2[6];        //str4 now holds 'o' as its value
```

Less than, equal to and greater than operator can be used to compare different string values. *valueOf()* method is used to convert the String Object to its corresponding primitive value. In order to append multiple string lines together or to store a very long string, two operators can be used, i.e. + or /.

```
let str = "This is an example to showcase a very long string"+
"and this string is not breaking, it is a continuous"+
"string and can be called via str";
let str2 = "This shows another way to store really long
strings \
      and using backslash allows it to span multiple \
      lines."
```

Static methods belonging to String Object include *String.raw()* that returns a newly created string using the raw template string. *String.fromCodePoint()* and *String.fromCharCode()* return a new string generated via the specified sequence of code points and Unicode values, respectively.

String Object has only one instance property, i.e. *String.prototype.length*, used to return the length of string. There are many instance methods that are defined in *String.prototype*, some of which are briefly described as follows:

- *at(index):* This method returns the character value present at the specified index. For negative values, reverse traversal is done.
- *charAt(index):* This method returns the character value present at the specified index value. For negative values, reverse traversal is done.
- *concat(s1,s2...):* This method returns a new string that is the concatenated or combined version of the passed arguments (two or more strings).
- *includes(str):* This method returns true or false after conducting a search to determine whether *str* exists in the calling object or not.
- *endsWith(str):* This method returns true if *str* is present at the end of the specified string; else it returns false.

- *indexOf(str) and lastIndexOf(str):* These methods return the index where *str* occurs first and last in the specified string, respectively. If *str* is not present, it returns −1.
- *match(regexp):* This method is used to determine the equivalence of a regex and a string. It returns an array for the match, if found; else it returns *null*.
- *matchAll(regexp):* This method returns an iterator for matches where each match corresponds to an entry in the array.
- *repeat(i):* This method returns a new string containing *i* number of copies concatenated together.
- *replace(X,Y) and replaceAll(X,Y):* This method is used to replace searched string, i.e. X, with the string Y within the calling string, where X can be a string or RegEx and Y can be string or a function. The former replaces first occurrence, and the latter replaces all the occurrences of X with Y.
- *search(regex):* This method searches the occurrence of RegEx within the calling string and returns the index value; otherwise, it returns −1.
- *slice(begin,end):* This method is used to extract a string starting from begin index to the end index. It returns the extracted string as a new string with no modifications on the calling string.
- *split(pattern):* This method is used to split a string using the *pattern* string into an ordered list of substrings, which are returned in an array format with every split string as an array entry.
- *startsWith(str):* This method returns true if the calling string begins with *str*; else it returns false.
- *substring(beg,end):* This method returns the substring extracted from the calling string where *beg* and *end* represent the starting and ending index value.
- *toLowerCase():* This method returns a new string by converting the calling string to a lowercase string.
- *toUpperCase():* This method returns a new string by converting the calling string to an uppercase string.
- *toString():* This method returns the string value for the calling object. It overrides the *toString()* method belonging to the Object prototype.
- *trim():* This method returns a new string by removing whitespace from both beginning and end of the string.
- *trimStart():* This method returns a new string by removing whitespace from the beginning of the calling string.
- *trimEnd():* This method returns a new string by removing whitespaces from the end of the calling string.

8.1.4.2 RegExp Object

Regular Expressions, or RegExp in short, are patterns that are used to find some combinations or patterns within strings [115]. They are composed of simple characters and modifiers. While searching a string for the presence of a RegExp, the match succeeds only when the exact RegExp is found in the exact order. Therefore, defining and declaring RegExp need careful attention to detail. A regular expression can be easily created either by using regex literal (enclosed within backslashes) or by calling constructor for *RegExp()* Object. This is shown in the example given next:

```
let a = /lookup/;            //using literal for regEx
//using constructor.
let b = new RegExp('lookup');
//string as first argument
let c = new RegExp(/lookup/);
                //literal as first argument
```

Creating RegExp objects via literal is comparatively faster if the expression doesn't change. If the RegExp is dynamic or changes constantly, then using the RegExp constructor is faster as it provides runtime compilation [116]. RegExp can also be created by using the following constructor where *pattern* represents a string value specifying the regular expression and the *flag* signifies an optional modifier string value containing any of the following flags:

- *g* flag specifies the global scope for the RegExp object, i.e. pattern searching will be performed within the global scope and searching will not stop after the first match.
- *i* flag specifies the case-sensitive nature of RegExp Object, i.e. when *i* flag is set, a case-insensitive searching is performed.
- *m* flag specifies whether the multiple matches across multiple lines are allowed or not.

Let us understand this with an example now. In the following example, we have created three regular expressions *a,b* and *c; a* is created using literals; *b* is created using constructor of *RegExp*, but the first argument is passed as a string; and *c* is also created using constructor of *RegExp*, but the first argument is passed as a literal. Further, we created a string *str* where we will apply regular expression to search for and replace the pattern.

```
let a = /lookup/i;            //using literal for regEx
//using constructor.
let b = new RegExp('lookup','i');
//string as first argument and modifier as second
```

```
let c = new RegExp(/lookup/,'i');
//literal as first argument and modifier as second
var str = "This is a string to perform lookup";
str.search(a);
str.replace(b,"Regular Expression Replacement");
```

In Figure 8.2, we can observe the output of the above code. The first output created is 28; this is the location of word "lookup" in *str*. The second output is the replacement operation using *RegExp* over *str*. The last three outputs demonstrate that no matter how we create *RegExp*, the expression created is same in all cases.

There are more features available in regular expressions where brackets are used to find a series or range of characters within strings. These bracket notations are demonstrated in Table 8.1. These can be used in the RegExp, and one can get a handle on them with practice and referencing the following Table 8.1.

```
> let a = /lookup/i;               //using literal for regEx
//using constructor.
let b=new RegExp('lookup','i');
//string as first argument and modifier as second
let c=new RegExp(/lookup/,'i');
//literal as first argument and modifier as second
var str="This is a string to perform lookup";
console.log(str.search(a));
str=str.replace(b,"Regular Expression Replacement");
console.log(str);
console.log(a);
console.log(b);
console.log(c);

28

This is a string to perform Regular Expression Replacement

/lookup/i

/lookup/i

/lookup/i
```

FIGURE 8.2
Output screenshot.

TABLE 8.1

RegExp Bracket Notation

Bracket Notation	Explanation
[abcd] or [0-9]	To find character(s) present between square brackets.
[^abcd] or [^0-9]	To find character(s) that are not present between square brackets.
(a | b | c | d)	To find any of the pattern/characters specified.

TABLE 8.2

Metacharacter in Regular Expressions

Metacharacter	Explanation
.	To search for a single character (excluding newline).
\w and \W	To find a *word* character and a *nonword* character, respectively.
\d and \D	To find a *digit* and *nondigit* character.
\s and \S	To find a *whitespace* and *nonwhitespace* characters, respectively.
\0	To find a *null* character.
\n	To find a *newline* character.
\r	To find a carriage return character.
\t and \v	To find a *tab* character and a *vertical tab* character, respectively.

To generate richer patterns in regular expression, certain *metacharacters* are offered by JavaScript. These metacharacters are characters with a special meaning and are summarized in Table 8.2.

Instance properties for RegExp Object include *RegExp.prototype.flags* that returns a string value containing flags present on the calling RegExp object. *RegExp.prototype.dotAll* is a read-only property used to show whether the *s* flag is used with the regular expression. It returns true if *g* flag was used; else it returns false. *RegExp.prototype.global* is also a read-only property used to indicate the state of *g* flag with the regular expression. It returns true if *g* flag was used; else it returns false. *RegExp.prototype.test()* is used to test for a pattern within the calling string. It returns true or false depending upon search result. *RegExp.prototype.toString()* returns the string value of the regular expression.

Regular Expressions are commonly used with Strings, especially the methods belonging to String Object including *math()*, *search()*, *split()* and *replace()*. Pattern matching, searching and replacing are common-use cases for regular expressions.

8.1.5 Keyed Collections

The Keyed Collections comprise objects that use keys to identify the values and are iterable in nature according to their respective order of insertion. These include *Set, Map, WeakSet* and *WeakMap* [117]. The following sub-sections discuss two of the most popular keyed collections used in JavaScript: Map and Set Object.

8.1.5.1 Map Object

Map Object can hold key value pairs in the exact same *order of insertion*. Primitive or Object value can be used either as a key or as a value. It easily iterates the entries via *for … of* loops where each iteration returns a key-value pair [118]. Map class provides a built-in constructor, i.e. *Map()*, to create new Map Objects.

Instance property for Map object is the *Map.prototype.size* property, which is used to return the number of key-value pairs present in the Map object. Instance methods defined in *Map.prototype* are listed next:

- *clear():* This method empties the Map object by clearing all the key-value pairs from it.
- *get(key):* This method returns the value corresponding to the specified key in calling object; else it returns *undefined*.
- *set(key, value):* This method is used to set the *value* with the specified *key* in the calling Map object. If a *value* already exists for the *key*, it is overwritten, and the resulting Map object is returned.
- *has(key):* This method returns a boolean primitive depending on the presence of a *key* within the calling object.
- *delete(key):* This method deletes the value corresponding to the *key* and returns true. If the *key* is not present, it returns false.
- *keys():* This method returns an iterator object comprising only *keys* (in order) for each element present in Map object.
- *values():* This method returns an iterator object comprising only *values* (in order) for each element present in the Map object.
- *entries():* This method returns an iterator object comprising an *array* *[key,values]* for the Map object in insertion order.

8.1.5.2 Set Object

Set object is simply a collection of *unique* items. These items can be either primitive values or object references. *NaN* and *undefined* can also be stored in a Set. Iteration is simply done on a Set object in the order of insertion [119]. Set class provides an in-built constructor *Set()* to create new objects.

Set class also has an instance property, i.e. *Set.prototype.size*, which returns the integer count of the number of values held in the calling Set object. Instance methods defined in *Set.prototype* are listed out as follows:

- *add(value):* This method is used to add a new value to the existing Set object, which is updated and then returned.
- *clear():* This method is used to remove all the values from the calling Set object.
- *delete(val):* This method deletes the passed value from the calling object and returns true; else it returns false.
- *has(val):* This method checks for the presence of *val* in the calling Set object and returns true if found; else it returns false.

- *values()*: This method returns an iterator object holding all the values in their insertion order for the calling Set object.
- *entries()*: This method returns a new iterator object that holds an *array[value, value]*, i.e. same value for both key and value, in insertion order.
- *forEach()*: This method executes a provided operation once for every value belonging to the calling Set object.

8.1.6 Indexed Collections

Indexed collections comprise objects such as *Array, Int8Array, Floar32Array, BigInt64Array, Int16Array* and so on. They represent ordered collections of data, which include arrays and array-like constructs [120]. The array object is discussed briefly in the following section as it was discussed at great length in Chapter 5.

8.1.6.1 Array Object

Array objects are the dynamic high-level objects with data stored at noncontiguous locations. Array prototype provides methods for easy traversal and manipulation operations on arrays.

Arrays in JavaScript are zero indexed with last element at (*length* -1) where *length* is the property associated with the array [121]. Using an invalid index resolves to *undefined*. Built-in array object includes methods like *slice(), indexOf(), join(), push() and splice()* to modify the length property of an array.

Array Object has its own constructor *Array()* to create an Array object. Static methods for Array include *Array.from(), Array.isArray()* and *Array.of()*. *Array.from()* is used to create a new Array object using the calling array. *Array.isArray()* is used to check whether the calling entity is an array or not and returns primitive Boolean value depending on the result. *Array.of(arg(s))* is used to create a new Array Object by using the passed arguments as array entries. Number of arguments, i.e. *arg(s)*, can vary while using *Array.of(arg(s))*.

```
Array.of(9);        //creates the Array - [9]
Array.of(4, 5, 6, 7, 8, 9, 10);
              //creates the Array - [4,5,6,7,8,9,10]
```

The only instance property for Array Object is the *length* property. Instance methods defined in Array.prototype are explained as follows:

- *at()*: This method returns the array item present at the specified index; negative index values are counted from the end of array.
- *concat()*: This method returns a new array after joining the other passed array(s) as arguments to *concat()*.

- *copyWithin()*: This method is used to copy a set of array items within the calling array.
- *entries()*: This method returns a new iterator object containing key-value pairs for the calling object.
- *every()*: This method returns true if the specified condition holds true for every element of the calling object.
- *fill()*: This method is used to fill the values of the entire array with the passed static value.
- *filter()*: This method tests the filtering condition on every element of the calling array and returns a new array comprising only those values that pass the filtering condition.
- *find()*: This method finds the array element that satisfies the specified condition. If found, true is returned; else *undefined* is returned.
- *findIndex()*: This method returns the index of the specified element within the array; else it returns −1.
- *forEach():* This method calls the specified method for every element present in the calling array.
- *includes(val)*: This method returns true if the calling array contains the specified value; else it returns false.
- *indexOf()*: This method returns the index of the first occurrence of the specified element in array; else it returns −1.
- *join()*: This method joins all the elements of the calling array and returns the resultant string.
- *keys()*: This method returns an iterator object containing index for every element of the calling array object.
- *lastIndexOf()*: This method returns the last index of the specified element in the array; else it returns −1.
- *pop()*: This method removes and returns the last element belonging to the calling array.
- *push():* This method adds a new element to the array and returns the length of the resultant array.
- *reverse()*: This method reverses the entire order of elements present inside the array and returns the resultant array object.
- *shift():* This method removes and returns the first element present within the calling array.
- *some()*: This method returns true even if a single element satisfies the specified condition.
- *sort()*: This method is used to sort the elements of the array and return the resultant array.

- *splice()*: This method is used to add or remove elements from the calling array object.
- *toString()*: This method returns a string containing all the elements belonging to the array.
- *values()*: This method returns a new iterator object containing values for all elements belonging to the array.

8.1.7 Structured Data

Structured Data includes the objects that are used to interact and represent the structured data and the data embedded within JSON (JavaScript Object Notation) [122].

8.1.8 Value Properties

Value properties are the global properties that return a simple, predefined value. They do not have any associated properties or methods. These include *Infinity, NaN, globalThis* and *undefined*.

- *Infinity* is a global property that represents infinity; this value can even be compared. The method *isFinite()* can be used to test whether a given object contains the infinity or not. For example:

```
let result = 12 / 0;                    //dividing by zero.
console.log(result);                    //stores Infinity
if (result = === Infinity)
    console.log('Oh this is infinity value')
console.log(isFinite(result));    //prints false
```

- *NaN* is a global property that represents Not a Number; this can be used during numerical computation where Not a Number can disrupt the output. The method *isNaN()* can be used to test any object that contains a number. For example:

```
let x = '1'                    //storing a number in x
console.log(isNaN(x))          //prints false
let y = 'a'                    //storing a string in y
console.log(isNaN(y))          //prints true
```

- *undefined* property represents undefined in JavaScript, which is primitive. This can be used at places where the value has not been defined. This value is returned when default values are not available for variables. For example:

```
let x
console.log(x);  // prints undefined
```

- *globalThis* property represents the reference of the global scope. JavaScript offers in different environments for different global scopes; this makes accessing global variables a tedious task. *globalThis* always has a set global reference, and this makes reference easier.

8.1.9 Function Properties

Function properties are the global functions that do not rely on any object to call; these respond directly to the caller. These include *isFinite()*, *isNan()*, *parseInt()*, *eval()*, *parseFloat()*, *encodeURI()*, *encodeURIComponent()*, *decodeURI()* and *decodeURIComponent()*.

- *eval():* This methods evaluates the JavaScript code that passed as a string. For example:

```
console.log(eval('12 + 24'));  //prints 36
```

- *isFinite():* This method checks whether the calling object resolves to *finite* or to *Infinite*. For example:

```
console.log(isFinite(10 / 0));  //prints false
```

- *isNan():* This method checks whether the calling object contains *NaN*. For example:

```
console.log(isNan('aa'));  //prints true
```

- *parseInt():* This method parses the given String to its corresponding Integer value based upon the second parameter that serves as radix value. For example:

```
console.log(parseInt('oxf', 16);  //prints 15
```

- *parseFloat():* This methods parses the given String to its corresponding Float value. For example:

```
console.log(parseFloat("3.14"));  //prints 3.14
```

- *encodeURI() and decodeURI():* These methods are used to encode or decode special character in a strings URL. These methods do not encode or decode these special characters (, /?: @ & = + $ #). For example:

```
var uritoEncode = "example.com?name = råvi";
var result = encodeURI(uritoEncode);
console.log(result);
// Output is "example.com?name = r%C3%A5vi"
var uriAfterDecode = decodeURI(result)
console.log(uriAfterDecode)
        //Output is example.com?name = råvi"
```

- *encodeURIComponent() and decodeURIComponent():* These methods are also used to encode or decode special character in a strings URL. These functions also encode or decode these special characters (, /?: @ & = + $ #). For example:

```
var uritoEncode = "example.com?name = råvi";
var result = encodeURIComponent(uritoEncode);
console.log(result);
// Output is ""example.com%3Fname%3Dr%C3%A5vi""
var uriAfterDecode = decodeURIComponent(result)
console.log(uriAfterDecode)
    //Output is example.com?name = råvi"
```

8.2 Exercise

8.2.1 Theory

i. Provide a classification for the built-in standard objects available in JavaScript. Elaborate and discuss each category.

ii. What is the difference between Object prototype and Function prototype? Do they provide the same functionality? If not, justify your answer.

iii. How can you use Number and Date objects while designing a web page? Critically brainstorm as to why we need these objects and their properties.

iv. List the properties provided by the Math object. Explain why most of these properties are static in nature.

v. What do you understand by Regular Expressions? Explain how RegExp can be used with Strings using suitable examples.

vi. Why do we need keyed collections in JavaScript? What is the difference between Map and Set objects? Provide the cases where one must use Map and Set objects, respectively.

8.2.2 True/False

i. Boolean object and boolean primitives return the same values as their result.

ii. Every symbol object inherits from the object prototype.

iii. Indexed collections allow unordered storage of elements.

iv. Value properties include NaN, Infinity, globalThis and undefined.

v. Using *window.prompt()* returns the primitive value true.

8.2.3 Multiple-Choice Questions

i. What is the format for using *Dates* in JavaScript?
   a. Number of days passed since January 1, 1970
   b. Number of milliseconds elapsed since January 1, 1970
   c. Number of seconds elapsed since January 1, 1970
   d. None of the above

ii. What is the correct method to create a new Date object?
   a. obj.Date(arguments);
   b. obj= new Date([arguments]);
   c. obj new Date();
   d. None of the above

iii. Chose the correct declaration for creating Symbol object:
   a. let symbolA = new Symbol('JavaScript');
   b. let symbolB = new Object(Symbol('JavaScript'));
   c. let symbolC = Symbol('JavaScript');
   d. All of the above

iv. JavaScript provides built-in objects because:
   a. it does not want the developer to design every new object from scratch.
   b. it works on prototypal inheritance that enhances its features.
   c. it provides a playground for newly created objects to inherit properties from.
   d. All of the above

v. The generic prototype from which most of the user-defined objects inherit from is _____ prototype.

   a. Function

   b. Array

   c. String

   d. Object

8.3 Demo and Hands-On for Loops

8.3.1 Objective

   i. Control flow via looping.

  ii. Counting the number of selected items in a form, with loop.

 iii. Traversing iterables with different loops.

8.3.2 Prerequisite

- **Step 1:** Visit https://www.routledge.com/9780367641429 and download zip file for *Ch8_Loops_Template.*
- **Step 2:** Unzip the content and open *index.html* in any editor of your choice (example: Notepad or Visual Studio Code).
- **Step 3:** Insert code given next inside **<script>** block in *index.html*.
- **Step 4:** Repeat the above three steps for each snippet given next.

8.3.3 Explore

Using any editor, we will write and test JavaScript code to use loops and see results on the web page. Here's what you should know about Loops. Loops allow us to control the flow of code and execute a specific code snippet until a condition evaluates true, or using range-based loops, which allow to conveniently traverse iterables like arrays, maps, etc.

8.3.3.1 Code Snippet-1

```
var p1 = "This is Placeholder for p1";
var p2 = "This is Placeholder for p2";
var p3 = "";
var p4 = "";
var p5 = "";
```

```javascript
document.addEventListener("DOMContentLoaded", () => {
    document.querySelector("#p1").innerHTML = p1;
    document.querySelector("#p2").innerHTML = p2;
    document.querySelector("#p3").innerHTML = p3;
    document.querySelector("#p4").innerHTML = p4;
    document.querySelector("#p5").innerHTML = p5;
});

function howMany(selectObject) {
    let numberSelected = 0;
    // For loop demo.
    for (let i = 0; i < selectObject.options.length; i + +) {
        if (selectObject.options[i].selected) {
            numberSelected + +;
        }
    }

    // While loop equivalent to the above for snippet.
    numberSelected = 0;
    let i = 0;
    while (i < selectObject.options.length) {
        if (selectObject.options[i].selected) {
            numberSelected + +;
        }
        i + +;
    }

    // Return result
    return numberSelected;
}

let btn = document.getElementById("btn");
btn.addEventListener("click", function () {
    alert(
        "Number of options selected: " +
howMany(document.selectForm.musicTypes));
});
```

8.3.3.2 Code Snippet-2

```javascript
// Utility ()
const appendToHTML = (contents, color = "white") => {
    contents.forEach((content) => {
        const myPara = document.createElement("p");
        myPara.style.color = color;
```

```
            myPara.innerHTML = content;
            resultHere.appendChild(myPara);
            sep();
        });
    };

    const sep = () => {
        const separator = document.createElement("hr");
        resultHere.appendChild(separator);
    };

/*********************************************************
***********************/

    // Reference to result screen.
    var resultHere = document.querySelector(".results");

    // Sample JSON Object.
    const godzilla = { type: "titan", size: "monstrous",
habitat: "hollow earth" };
    // Print JSON in webpage
    appendToHTML([
        "JSON message hardcoded = >",
        JSON.stringify(godzilla),
        "Iterating JSON object via `for… in`",
], 'black');

    // Using { FOR… IN } loop to iterate through object.
    for (let i in godzilla) {
        appendToHTML([i + ": " + godzilla[i]]);
    }
```

8.3.3.3 Code Snippet-3

```
    // Utility ()
    const appendToHTML = (contents, color = "white") => {
        contents.forEach((content) => {
            const myPara = document.createElement("p");
            myPara.style.color = color;
            myPara.innerHTML = content;
            resultHere.appendChild(myPara);
            sep();
        });
    };
```

```javascript
    const sep = () => {
        const separator = document.createElement("hr");
        resultHere.appendChild(separator);
    };
/*********************************************************
**********************/

    // Reference to result screen.
    var resultHere = document.querySelector(".results");

    // Sample Array.
    const favMovies = [
        "Interstellar",
        "Tenet",
        "John Wick",
        "Harry Porter",
        "How to train your dragon",
        "Jodha Akbar",
    ];

    // Print JSON in webpage
    appendToHTML(
        [
            "Array FavMovies = >",
            "[" + favMovies.toString() + "]",
            "Iterating JSON object via `for… each`",
        ],
        "yellow"
    );

    // for… each
    favMovies.forEach((element) => {
        appendToHTML([element], "black");
    });

    // Insert separator
    sep();

    appendToHTML(
        [
            "Array FavMovies = >",
            "[" + favMovies.toString() + "]",
            "Iterating JSON object via `for… of`",
            "For… of statement is recommended for traversing
```

```
iterables like arrays, instead of for… in as for…in also
iterates over user-defined properties in arrays.",
    ],
    "yellow"
);

// for… of
for (let content of favMovies) {
    appendToHTML([content], "black");
}
```

8.4 Demo and Hands-On for Hoisting

8.4.1 Objective

  i. Understanding hoisting and its significance.
 ii. Hoisting of functions.
iii. Hosting of variables.

8.4.2 Prerequisite

- **Step 1:** Visit https://www.routledge.com/9780367641429 and download zip file for *Ch8_Hoisting_Template.*
- **Step 2:** Unzip the content and open *index.html* in any editor of your choice (example: Notepad or Visual Studio Code).
- **Step 3:** Insert code given next inside <**script**> block in *index.html*.
- **Step 4:** Repeat the above three steps for each snippet given next.

8.4.3 Explore

Using any editor, we will write and test JavaScript code to interact with the DOM and see the results on the web page. Here's what you should know about Hoisting. Conceptually, in Js it's believed that all declarations (variables) and function definitions are moved to the top of the code, and hence, Js does not give reference error (or is aware of the variables), but that's not what happens. Variable declarations are right where they're defined, but Js stores them in the memory during compilation and hence is aware of those during runtime. This concept is called hoisting.

8.4.3.1 Code Snippet-1

```javascript
// Utility ()
const appendToHTML = (contents, color = "white") => {
    contents.forEach((content) => {
        const myPara = document.createElement("p");
        myPara.style.color = color;
        myPara.innerHTML = content;
        resultHere.appendChild(myPara);
        sep();
    });
};

const sep = () => {
    const separator = document.createElement("hr");
    resultHere.appendChild(separator);
};

/*****************************************************
********* */
var resultHere = document.querySelector(".results");

appendToHTML([
    "Hoisting is a context how javascript code execution
works.",
    "All functions and variable declarations are stored in
memory during compilation, and JS is aware of them. Example
-> ",
]);

// testFunction(x) isn'tdefined yet in the code above, but
will work fine as it will be stored in memory
// during compilation and Js will be aware about it ar
runtime.
testFunction("Js is so good!");

// Definition.
function testFunction(arg) {
    appendToHTML([
        `Hello I'm working because of HOISTING, yayyy |
Argument passed -> ${arg}`,
    ]);
}
```

8.4.3.2 Code Snippet-2

```javascript
// Utility ()
const appendToHTML = (contents, color = "white") => {
    contents.forEach((content) => {
        const myPara = document.createElement("p");
        myPara.style.color = color;
        myPara.innerHTML = content;
        resultHere.appendChild(myPara);
        sep();
    });
};

const sep = () => {
    const separator = document.createElement("hr");
    resultHere.appendChild(separator);
};

/*****************************************************
********* */
var resultHere = document.querySelector(".results");

appendToHTML(
    ["Hosting Does not work on initialization, only
declaration!"],
    "yellow"
);

// This works fine, variables declared & initialized and
then used in code.
var num; // Declaration
num = 6; // Initialization.
appendToHTML([`Variable num: ${num}`], "lime");

// Using variable before declaration.

// Let's declare with 'const' and 'let'***UNCOMMENT THIS
SNIPPET AND CHECK THE REFERENCE ERROR, JS IS NOT EVEN AWARE
ABOUT THESE VARIABLES HENCE const AND let is not hoisted******
*****************

// appendToHTML([
//      `Variable constVariable: ${constVariable}`,
//      `Variable letVariable: ${letVariable}`,
//]);
```

```
    // const constVariable;
    // let letVariable;
    // constVariable = "Helu";
    // letVariable = "World";

    /******************************************************
************/

    sep();
    //Let's try with 'var' and see how it goes.
    appendToHTML([`Variable thatsWeird: ${thatsWeird}`]);
    var thatsWeird; // Declaration.
    thatsWeird = 10;
    appendToHTML(
        [
            "It's undefined, because only DECLARATIONS ARE
HOISTED, NOT INITIALISATION.",
            "Atleast code is successfully executed, Js is aware
about the variables now."
        ],
        "tomato"
    );
    // Definition.
    function testFunction(arg) {
        appendToHTML([
            `Hello I'm working because of HOISTING, yayyy |
Arguement passed -> ${arg}`,
        ]);
    }
```

9

Validation, Error Handling and Cookies

Look for the exception to every rule and you'll find it.

–Marty Rubin

JavaScript is a dynamic and flexible language, just like we have discussed and demonstrated using different programs throughout the book. Many a times, this flexibility leads to creation of complex relationships between objects and other operating elements. Presence of a flaw within a program is called a *bug*. The process of finding and removing these bugs from program code is called *debugging*.

These bugs can arise due to *syntactical* or *logical* problems in code. Most important, they can also arise at *runtime* while executing code. With JavaScript, the possibility of runtime problems is very high due to the nature of the language itself. For developers, it is important that the web page continues to work under various circumstances. This is done by carefully designing and writing program code in order to remove syntactical and logical errors. However, for runtime errors, this is not possible because they will arise when the program is in motion and is executed at the user end, which can lead to a varied range of circumstances due to presence of heterogeneous browser environments. For JavaScript, objects pass through multiple interfaces inheriting properties and methods making it a rich yet a complex landscape for errors to arise.

It is important for every programming language to manage errors and other problems arising with code execution. JavaScript offers error handling to handle such problems. In this chapter, we discuss program code validation followed by the use of try, catch and finally blocks. We then discuss the use of finally block and throws keyword to generate errors. Error objects with their properties are also discussed. Lastly, we discuss the use and operations on cookies in detail.

9.1 Validation

While working with web pages, there are usually two cases while interacting with end user. Either we display information to the user or we take information from the user. Today, most of the web pages are interactive in

DOI: 10.1201/9781003122364-9

nature and require custom user input for working differently under different use cases.

For taking user input, HTML forms are the most commonly used alternative. These forms specify the input fields and make a beautiful interface to take input from users. They need to have very clear, concise and specific information about what they need from end user. Every form field has some associated data that needs to be in a proper format for further processing. *Validation* refers to the process of making sure that information passed by the user is correct and adheres to the specified program needs. Information validation can be done on either the client side or the server side. This is discussed in detail in the following section.

9.1.1 Server- and Client-Side Validation

Information validation can be performed on the client side where the user can be restricted to provide information in particular format. It has now become a standard filter mechanism where information is validated at user end. Another way to validate information is to carry out server-side validation where the input sent by the user is analyzed at server end.

Server-side validation is slower as the user input will need to go to server and come back with response. This creates delay in information processing at user end. Keeping both of these validation approaches in mind, it is recommended to use both client-side as well as server-side validation for user input.

Client-side validation is helpful as a filtering mechanism at the user's end [1]. It helps in enforcing certain constraints on the user such as choosing an appropriate password, making fields mandatory to fill, specifying format for data entry etc. Client-side validation can greatly reduce server overhead and make information processing easier.

On the other hand, it is also recommended to have server-side validation along with client-side validation. Even after enforcing client-side validation, users can behave in a malicious way toward the server side. This can be solved by applying strict server-side information validation on the input received from the user.

Validation is mostly performed on forms as they are the de facto standard for taking input or data from end users. HTML forms are the means to receive detailed input from the user. For validation purposes, both JavaScript and HTML can be used. HTML offers built-in form validation, whereas JavaScript can be used to enforce it with manual code. Both of these are discussed in subsequent sections.

9.1.2 Using Built-In Form Validation

HTML5 offers built-in form controls that greatly help in validation. Using these features, user data can be validated at client side [2]. Some of the form validation attributes provided by form elements are listed as follows:

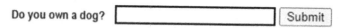

FIGURE 9.1
Expected output.

- *minlength* and *maxlength*: Represent the minimum and maximum length of text input (string), respectively.
- *min* and *max*: Represent the minimum and maximum value for a numeric input value, respectively.
- *required*: Indicates whether the form attribute is mandatory to be filled by user or not. Represented by a red-colored asterisk (*) at the top of form element.
- *type*: Indicates the type of input that is permitted on a form element such as a text, string and email.
- *pattern*: Represents a regular expression that defines the exact pattern a user input has to follow to enter the form details.

Using such form validation attributes allows only validated information to reach the server side. When the form element value is valid, it is allowed to submit (if JavaScript code is not blocking it). When the form element value is invalid, the browser blocks the form and displays the corresponding error message. Following program code(s) demonstrates the use of HTML built-in form validation features, followed by the screenshot of the output.

- The code without HTML validation:

```
<form><!--first entry for form-->
    <label for = "dog"> Do you own a dog? </label>
    <input id = "dog" name = "dog"> //not mandatory to fill
    <button>Submit</button>
</form>
```

The following screenshot (Figure 9.1) demonstrates the output for the above code; this form will not do any validation, i.e. even if no input is supplied. This form entry is not *required*, and thus, a user can directly submit it and it will be sent directly to server.

- The code with HTML validation:

```
<form><!--second entry for form-->
    <label for = "cat"> Do you own a cat? </label>
    <input id ="cat" name = "cat" required minlength = "2"
maxlength = "20"> <!--makes it mandatory to fill with length -->
```

```
constraints
<button>Submit</button>
</form>
```

For this form, the attribute inside *input* tag is *required*, which makes it mandatory for the user to enter. If a user submits the form without entering a value for this, the browser blocks the form and does not send it to server. If the user enters a value and it does not meet the criteria set by *minlength* and *maxlength*, the form is blocked and not sent.

Every user input must meet the criteria set for validation by the form. The following screenshot (Figure 9.2) demonstrates the output for the above code; this form will perform *minlength* and *maxlength* HTML validation, i.e. user will have to enter a string with at least 2 characters with a maximum limit of 20 characters.

HTML5 form validation is fairly static in nature and does not allow custom modifications. It supports the provided attributes and enforces validation to that extent only with a little scope of modification. It does not allow setting up of complex constraints on user values. However, HTML5 form validation is comparatively faster with respect to performance. JavaScript validation can solve these problems, and it can provide more complex validation on user input. Let us learn about it in the following section.

9.1.3 Using JavaScript for Validation

Validation can also be achieved by using JavaScript. Most of the browsers today support the usage of *Constraint Validation API*, which checks the user-entered values before passing them to server [3]. It can be used along with HTML form validation. Once the user input values are validated by HTML form elements, it is then passed to JavaScript code for further validation.

JavaScript then uses Constraint validation API to specify complex constraints on user input. This can be done on individual form elements or on the entire form itself, i.e. *<form>* element. It provides the following properties that are useful for validation:

- *validity*: This property returns a *ValidityState* object containing all the validation errors as the read-only properties of object for the given form element. Properties for *ValidityState* object include *tooLong, tooShort, rangeOverflow, patternMismatch, valueMissing* etc.

Do you own a cat? (required) * ⎣_____⎦   ⎣ Submit ⎦

FIGURE 9.2
Expected output.

- *validationMessage*: This is a read-only property used to retrieve the validation message to be displayed if the value of an element is invalid. If the form value is valid, it returns an empty string.
- *willValidate*: This property returns a read-only Boolean property specifying whether the given element is used for constraint validation or not.

Following methods are provided by the Constraint Validation API for validation of form elements:

- *checkValidity()*: This method returns a Boolean value depending on the value of element with respect to the specified constraints. For invalid value, it returns *false*; else it returns *true*.
- *reportValidity()*: This method is used to check whether an element's value meets the constraint needs or not. If value is invalid, it fires an *invalid* event, returns *false* and informs the user about the validity status.
- *setCustomValidity(msg)*: This method is used to set a custom message that will be displayed to the user when a given value is invalid or not acceptable. It is useful for specifying the constraints to user via an error message.

Constraint Validation API also provides extension to other interfaces for form-related validation. Such extension allows addition of additional properties and methods. Some of these interfaces include *HTMLSelectElement (<select>)*, *HTMLInputElement (<input>)*, *HTMLOutputElement (<output>)*, *HTML ButtonElement (<button>)* etc.

One example is provided at the end of this chapter to observe validation in action using JavaScript.

9.2 Error Handling

Apart from user input or form validation, there can be other errors within a program. Syntactical and logical errors are handled while designing the program code. They are fairly easy to handle. *Runtime errors* are the errors that occur when a program is under execution. These are the most important ones and thus most difficult to remove, especially while working with JavaScript.

Features of JavaScript such as the use of prototypes, inheritance, type casting, objects etc. make it even harder to identify errors as they may not arise instantly and take a significant time of processing under different cases to manifest. They take a lot of effort to find and correct. The mechanism to overcome these situations is known as *error handling*. The JavaScript runtime creates a special *error* object and *throws* it back to the caller, which is the web page or console, by default.

JavaScript also provides mechanism to catch this thrown object and look for alternative solutions to resolve the error in runtime. This is done using the *try*, *catch* and *finally* blocks. Let us understand the working of *error* object and *try*, *catch* and *finally* blocks in upcoming sections.

9.2.1 Error Object

JavaScript offers an in-built object that is used to provide information about the errors that are generated during program execution [4]. Error object provides such properties that give information about the error.

Two of the most commonly used Error object properties are *name* and *message*. The *name* property for error object is used to either set or return the error name. The *message* property is used to either set or return an error message in the form of a string.

Name property for error object can hold different values where each value denotes a distinct type of error. These values are explained as follows:

- *Range Error:* A range error indicates that the number value being provided falls outside the range of legal/acceptable values.
- *Reference Error:* A reference error is thrown when a reference is used without declaring it beforehand.
- *Syntax Error:* A syntax error represents the incorrect use of syntax in the code.
- *Type Error:* A type error is thrown when the used value falls outside the range of expected types such as using string in place of a number or vice versa.
- *URIError:* A Uniform Resource Identifier error is thrown when illegal values are used to define a URI value.

Error object has its own constructor to initialize the error object. Using the constructor *Error()* creates a new error object. Error object has the *toString()* instance method associated with it. It overrides the *Obejct.prototype.toString ()* method to return a string indicating the details about the calling Error object.

```
try {
    throw new Error("Just for fun")
}
catch (err) {
```

```
    console.log("Error details are: [Name:"+err.name +
"] and Message: +err.message + "]");
    }
```

9.2.2 Try-Catch Block

The *try* and *catch* blocks are the means to provide error handling feature in JavaScript. The program code that can possibly generate error and stop the execution of code can be placed inside the *try* block [5]. When the code within a *try* block is executed, there can be only two possibilities, i.e. either it generates an error or it does not.

Whenever an error occurs, the *try* block will pass the execution to the corresponding *catch* block. The code in the *catch* block is then executed. On the other hand, if the code within a *try* block is executed and no error occurs, then the corresponding *catch* block is not executed. Use of *try* and *catch* blocks makes the process of error handling a lot easier.

This way a testing mechanism is enforced that allows the program code to execute safely inside a *try* block. If in case, any error arises, the *catch* block is present to handle it. These blocks are always used together, i.e. for every *try* block, there will be a *catch* block. These blocks can be nested within one another to handle more complex cases. While nesting blocks, *try* block is paired with the closest *catch* block. One *try* block can have multiple *catch* blocks. Syntax for using these blocks is provided as follows:

```
Syntax:
try
{
        //code to be executed
}
catch(error)
{
    //code to handle the passed error
}
```

There can be multiple *catch* blocks for a given *try* block. A *try* block can contain different program statements that can throw different errors. For every error, a different catch block can exist. A small code segment demonstrating the use of *try-catch* blocks is given as follows and Figure 9.3 demonstarted the output.

```
<html>
<body>
  <script>
      try {
          let x = "Hello World!";
```

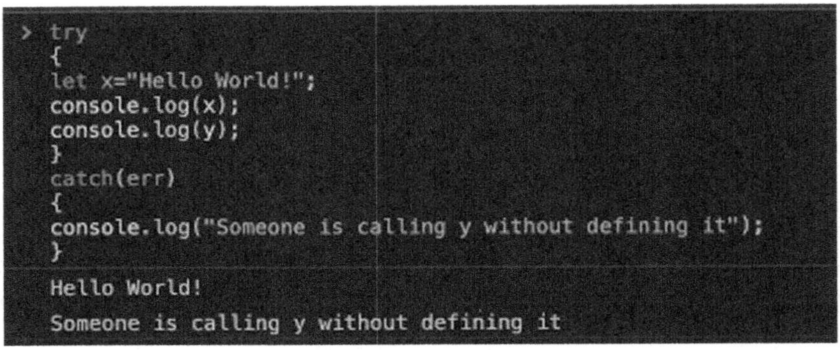

FIGURE 9.3
Output in web browser console.

```
        console.log(x);
        console.log(y);
    }
    catch (err) {
        console.log("Someone is calling y without de-
fining it");
    }
  </script>
</body>
</html>
}
```

9.2.3 Finally Block

The *finally* is an optional block of code that is provided with the *try-catch* blocks. It is most often used to perform cleanup operations once program execution is over such as closing a file, deleting a resource, displaying a message to user etc. Whether an exception is thrown or not thrown, statements inside a *finally* block are always executed exactly once. Syntax for using *finally* with the *try-catch* blocks is provided next:

```
Syntax:

try
{
      //code to be executed
}
catch(err)
{
      //catching the error
```

```
}
finally
{
// This code always executes once, before program execution
is complete.
}
```

9.2.4 Throw Statement

The *throw* statement is used to throw a custom user-defined error within a *try* block. Any user can define its own exception and throw it using the *throw* keyword. This is helpful for developers to identify errors in a unique and customized way [6]. Any program statement after the *throw* keyword is not executed, and the control passes directly to the corresponding *catch* block.

If a corresponding *catch* block does not exist, the program terminates. Syntax for using *throws* inside *try-catch* blocks is provided as follows:

```
Syntax:

try
{
      //code to be executed;
      throw userDefinedError;         //throw statement
}
catch(err)
{
      //code for handling that particular exception
}
```

throw statements are useful to generate exceptions that are understandable by the user and are applicable within the context of program code.

9.3 Cookies

Traditionally, once a server sent a web page to a user, it completely forgot about the user, and a fresh page was served every time the user request arrived at the server. A user's state and behavior information were not maintained, and this led to repetitive information disclosure from user end etc. This also led to wastage of precious bandwidth, time and resources due to recurrent loading of same information. To solve this problem, small text files were stored on the user's computer in the browser's file system to represent the state of an end user. These small files are called *cookies*.

Cookies are primarily key-value pairs used to store information about the user. These small text files are sent along with the user's request for a web page to the web server [7]. Using cookies, it becomes easier for the server to remember the state and information for a particular user.

document object for every web page has a unique property for storing the value of cookies. *document.cookie* property is used to read and write cookies associated with a given document. Using this property, cookies can be accessed as well as set to a particular value.

```
let cookieALL = document.cookie;
//returns a string containing all cookies in key-value pairs
document.cookie = newKeyValue;
//to add a new value to cookies
```

By default, cookies are destroyed once the browser closes. Such cookies are called as *session cookies*. In order to make a cookie survive, either the *expires* or the *max-age* property on the cookie has to be set. The parameter values are passed in milliseconds.

```
document.cookie = "user = Alex; max - age = 0";
//to remove instantly
document.cookie = "user = Alex; max - age = 3600";
//deletes after 1 hour
```

If the *secure* attribute is set in a cookie, then it must be transferred only using the HTTPS protocol to enable secure transfer of information.

```
document.cookie = "user = Ravi; Secure"; //sent over HTTPS
```

9.4 Strict Mode

Strict mode is a new addition to the ECMAScript5 specification, which allows the program and functions to execute in a strict context. *use strict* is a literal expression that is used to prevent the program code from certain actions [8]. It silently removes some of the subtle errors and directly throws them on the console.

It eliminates silent errors and restricts use of certain utilities, which may be defined in future version of the language. Using strict mode, a program executes faster and becomes more optimized for search engines. It automatically disables the use of features that are incomplete, confusing or not secure enough.

When a script is not operating in *strict mode,* it is then called *sloppy* in nature. Strict mode can be applied either globally to the entire script or to individual functions. A program code can be instructed to use the strict mode by writing *use strict* at the top of the script or before the function declaration. Writing *use strict* anywhere else will not lead to using of strict mode in the code.

```
'use strict'
console.log ("A stricter script in action");
```

Another example to understand *strict* mode; let's see a code snippet without using strict mode and its output.

```
var myVariable; //Creating myVariable object
myVaroable = 10;    //Observe the misspelled myVaroable
console.log(myVaroable);    //trying to print misspelled
console.log(myVariable);    //trying to see correct one
```

Now observe the output (Figure 9.4) of the above code; there is no impact because of misspelled object names. This introduces unintentional errors in the code.

Now, let us see the same code using the *use strict* keyword placed at top of the code:

```
'use strict';           //using strict mode

var myVariable; //Creating myVariable object
myVaroable = 10;        //Observe the misspelled myVaroable
console.log(myVaroable);    //trying to print misspelled
console.log(myVariable);    //trying to see correct one
```

Carefully observe the output (Figure 9.5) now; the runtime environment recognized the misspelled word and throws a *ReferenceError.* This is how strict mode helps in creating a stable code and resulting in an optimized and safer code. Use of strict mode is recommended for faster, safer and optimized execution of program code.

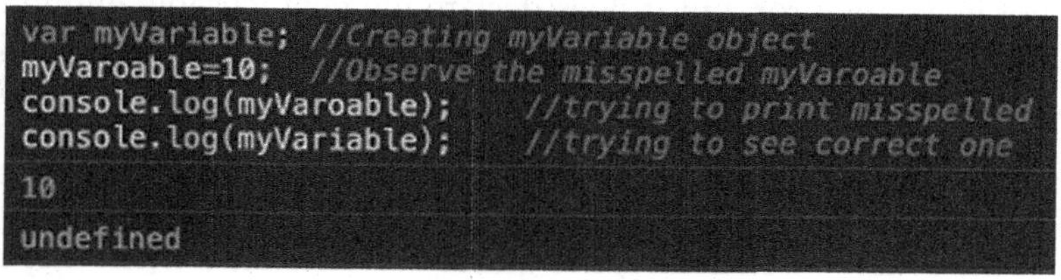

FIGURE 9.4
Output screenshot.

```
> 'use strict';          //using strict mode

  var myVariable; //Creating myVariable object
  myVaroable=10;          //Observe the misspelled myVaroable
  console.log(myVaroable);    //trying to print misspelled
  console.log(myVariable);    //trying to see correct one
⊗ ▶Uncaught ReferenceError: myVaroable is not defined
    at <anonymous>:4:11
```

FIGURE 9.5
Output screenshot.

9.5 Best Practices

While writing your JavaScript code, you should ideally follow some best practices in order to avoid any unnecessary roadblocks, debugging and delay. By adhering to the best practices outlined next, a developer reduces the chance of faulty code, makes it easier for others to understand and the end result is a clean, efficient and meaningful code, which can be easily upgraded as and when needed. Some of these practices are listed as follows:

- Always use comments to explain your code. Program code is written once, but it is read multiple times over its lifetime. It is a best practice to properly comment your code by explaining whatever is necessary. It should be generic and understandable to anyone who operates on the code at later point in time.

- Use meaningful variable and function names. Instead of naming your variables as *x, y, z*, it is better to name them as *noOfStudents*, *calculateSum()*, *printTotal()*. Proper naming of variables and functions makes it fairly easier to understand the working of code, especially in cases where variables and functions are passed in complex relationships throughout the life cycle of a program.

- Avoid the use of global variables; use *var* wherever possible. Skipping the *var* keyword and directly using it in program code do not generate error, but it does declare the new variable within the global scope and then uses it. Creating multiple variables in global scope is not considered a good practice as it increases the global footprint of your code. This increases the possibility of those variables to mistakenly get exploited by other applications, libraries etc. It also leads to unnecessary collision and decreases the performance of the program. Therefore, use global variables only when they are specifically needed; else use the *var* keyword to declare them within a given scope.

- Not every browser supports JavaScript. It sometimes happens that the user browser has disabled the use of JavaScript. Under such circumstances, it is best to design a web page with only HTML code first, to make sure that a meaningful page is displayed to user and not a broken script of code. Once a basic HTML page is properly worked on, a developer can then focus on integrating JavaScript in the web page to enhance its functionality.

- Use literals to create arrays, objects etc. instead of the *new* constructor. When an array (or function or object) is created using the constructor, it inherits all properties and methods associated with its prototype. For simple operations like handling information in an array, it is not mandatory to make an array object. Instead, a simple array would just work fine and increase the operating speed. On the other hand, creating an array object for the same will not lead to any considerable advantage. Rather it would turn out to be comparatively slower. It is thus recommended to use object literals to initialize basic objects instead of the constructor initialization.

```
var obj = {}              //using literal
var obj2 = new Object();    //new object
var arr = [];             //simple array
var arr2 = new Array();   //Array object
```

- Omit the repetitively used *var* keyword, and use commas to declare a list of variables. It is not mandatory to declare every variable in a separate line. Instead, a comma-separated list can be used with the *var* keyword. This is an optimized way to work with variables. Try to declare and initialize the variables at the beginning of your script instead of declaring them anywhere throughout the program. Set default values for variables whenever you declare them.

```
var noOfItems;
var userID;
var nameOfDog;
var noOfItems, userID, nameOfDog;    //works the same
```

- Browsers automatically insert a semicolon at the end of program statements. However, solely relying on the browser to finish the statements can lead to a problem later on. It is advised to properly terminate lines using semicolons wherever applicable.

- Use of raw JavaScript to execute something is always faster than using a library to do the same. Using a library enhances the functionality of JavaScript but delays the operating speed. Using raw JavaScript can execute it swiftly, but writing the code is difficult than using a predefined library. A careful decision, therefore, must be made regarding the balance of operating speed and use of extended functionality for the web page. Often, the decision changes for different web pages or use cases.

- Placing scripts at the bottom of the HTML page allows faster loading of the web page. JavaScript's *<script>* tag is used to provide enhanced user behavior by introducing dynamic features such as behavior *onClick()*, *onLoad()* etc. This script can safely execute once the HTML is loaded.

- For debugging purposes, use JSLint debugger that helps the developer to find the problem in JavaScript code and reports the source for it. It parses the code to find syntax errors, logical errors or structural errors in code and reports them back to the developer.

- Use '===' instead of '==' (or use '==!' instead of '=!'). The operator '===' compares the value as well as the type for the passed arguments, whereas the operator '==' first performs type conversion and then compares the value. Therefore, wherever needed, use the '===' operator for faster and more meaningful comparison.

- Use the *strict mode* to avoid mistakes, and adhere to a stricter and clean coding practice. Using this mode disables the use of *with* and other operations that lead to unnecessary complexity in code.

- Try to design and program in a modular and specialized manner. It is often tempting to use flexible nature of JavaScript to carry out multiple tasks using a limited number of functions. However, this should be avoided, and only specialized functions for specific tasks must be used. This helps in easy reuse, debugging and extension of code without the need of rewriting.

9.6 Exercise

9.6.1 Theory

i. Write a script for validating different form fields where the fields must have a Boolean, a Number and a String value in that order with maximum length of 15.

ii. Using JavaScript code, differentiate how client-side form validation is achieved using HTML and using JavaScript's Constraint Validation API.

iii. While working with the text of this book, you must have practiced a lot of code with the exercises given at the end of this chapter.

Using your practice code and experience, list at least 10 of the best practices that you think work well for you.

iv. What are the different Error name values in JavaScript?

v. List out the different types of errors used in JavaScript. Explain each of them with suitable examples.

vi. Differentiate between client-side and server-side validation. According to you, how validation should be handled? Elaborate on your answer.

9.6.2 True/False

i. Error notifications are shown in the console for both syntax and runtime error.

ii. A bug is defined as any mistake that is present inside a script.

iii. Boolean function returns *true* for anything which is nonempty.

9.6.3 Multiple-Choice Questions

i. Form validation happens at:
   a. client side.
   b. server side.
   c. Both
   d. None of the above

ii. Which of the following functions is called to validate data entered by the user?
   a. onValid()
   b. validate()
   c. checkData()
   d. No predefined function has to be called

iii. What is the output for the following program code?

```
function myFunction() {
   var y = " ";
   document.getElementById("demo").innerHTML = Boolean(y);
}
```

   a. False
   b. True
   c. RangeError
   d. None of the above

9.7 Demo and Hands-On for Form Validation and Cookies

9.7.1 Objective

i. Apply custom rules of validation to form.

ii. Let these rules be middleware checked, before allowing submit.

iii. Store form details into cookies upon successful validation.

9.7.2 Prerequisite

- **Step 1:** Visit https://www.routledge.com/9780367641429 and download zip file for *Ch9_ValidationCookie_Template*.
- **Step 2:** Unzip the content and open *index.html* in any editor of your choice (example: Notepad or Visual Studio Code).
- **Step 3:** Insert code given next inside <**script**> block in *index.html*.
- **Step 4:** Repeat the above three steps for each snippet given next.

9.7.3 Explore

Using any editor, we will write and test JavaScript code to use form validation and store the form details into cookies.

9.7.3.1 Code Snippet-1

```
// Utility ()
const appendToHTML = (contents, color = "white") => {
    contents.forEach((content) => {
        const myPara = document.createElement("p");
        myPara.style.color = color;
        myPara.innerHTML = content;
        resultHere.appendChild(myPara);
        sep();
    });
};

const sep = () => {
    const separator = document.createElement("hr");
    resultHere.appendChild(separator);
};

/****************************************************
********** */
    var resultHere = document.querySelector(".results");
```

```
    // DOM references
    const nameF = document.getElementById("name");
    const emailF = document.getElementById("email");
    const pwdF = document.getElementById("pwd");

    const btn = document.getElementById("btn");
    const getC = document.getElementById("get");

    btn.addEventListener("click", (e) => {
        // TODO: Validate name
        // Validating Name, name can be any weird
AudioWorkletNode, let's just make sure it's not empty.
        if (!nameF.value.trim()) {
            alert("Name cannot be empty:/");
            return;
        }

        // TODO: Validate email
        // Let's say we only allow emails from gmail.com, not
any other domain, so...
        console.log(emailF.value.trim().toLowerCase().
split("@")[1]);
        if (emailF.value.trim().toLowerCase().split("@")
[1] != "gmail.com") {
            alert("Sorry, only gmail accepted, go make a new
account.");
            return;
        }

        // TODO: Validate Password.
        // Must be 8<= character-count<= 15 characters, must
have english alphabets only.
        const pwd = pwdF.value;
        if (8<= pwd.length<= 15) {
            // Regex Expression for alphabets
            var alpha = /^[a-zA-Z] + $/;

            // Checking for matches
            if (!pwd.match(alpha)) {
                alert("Password must contain alphabets
only<3");
                return;
            }
        } else {
```

```
        alert ("Must be 8<= character-count<= 15
characters");
        return;
    }

    // If all checks are passed, control reaches here, now
store the form data in cookie.
    document.getElementById("res").style.display =
"block";
    setCookie(pwdF.value, `${nameF.value.trim()}-
${emailF.value.trim()}`);
  });

  // Get cookie and display in console.
  getC.addEventListener("click", (e) => {
    console.log(getCookie(pwdF.value));
  });
  /****************************************************
  ********** */
  // Cookie Setter
  function setCookie(cname, cvalue) {
    var d = new Date();
    document.cookie = cname + "=" + cvalue + ";";
  }
  // Cookie Getter
  function getCookie(cname) {
    var name = cname + "=";
    var ca = document.cookie.split(";");
    for (var i = 0; i<ca.length; i++) {
      var c = ca[i];
      while (c.charAt(0) == " ") {
        c = c.substring(1);
      }
      if (c.indexOf(name) == 0) {
        return c.substring(name.length, c.length);
      }
    }
    return "*";
  }
```

10

Miscellaneous Case Studies

Even if yesterday was wildly successful, I still don't want to repeat it. Rather, I want to build on it.

−Craig D. Lounsbrough

Our journey of learning JavaScript is finally coming to an end. It may be the end of learning basic JavaScript language know-how, but there is so much more that we can still learn. In this chapter, we aim to give you a peek into some of the more powerful features of JavaScript. This chapter provides a total hands-on practice-based exercise for you to get a taste of some other topics. This involves three comprehensive case studies that are meant for you to explore and then implement your own learning. These case studies include few topics that are beyond the scope of this book, but this will give you an overview of how JavaScript can be taken to the next level and how JavaScript helps in designing various other applications. The scope of JavaScript is very vast. Our book was just one of the first stops to set the playground for you to fly. We wish you all the success and fulfilling joy while programming with this language.

10.1 Case Study-1: Introduction to Fetch API

10.1.1 Objective

   i. Learning about the use of Fetch API used by JavaScript.
   ii. Calling an API using *fetch()*.
   iii. Parsing the data from the API.

10.1.2 Prerequisite

   i. Visit https://www.routledge.com/9780367641429 and download zip file for *Ch_10_Fetch_Template*.
   ii. Unzip the content and open *index.html* in any editor of your choice (example: Notepad or Visual Studio Code).
   iii. Insert code given next inside **<script>** block in *index. html*.

DOI: 10.1201/9781003122364-10

iv. Repeat the above three steps for each snippet given next.

10.1.3 Explore

Using any editor, we will write and test JavaScript code to fetch API and see the results on the web page. Here is what you should know about Fetch. Fetch is an easy, logical way to GET and POST resources across networks [1]. It provides a JavaScript interface used for manipulation of resources. We're just looking into the GET aspect of *fetch()* syntax.

- The API we are going to use for this case study is Dog-API. We recommend reading its documentation before proceeding with the exercise [2].
- After executing this case study and practicing with it, you can further explore more JavaScript APIs, which would be fun to learn [3].

```
fetch(<API' s url>)
// fetch will take some time because it's getting
resources from a different server, hence the next step must
be executed when the API has been successfully fetched.
  .then(<res-variables>=> { return res.json() }
  .then(<data-variable>=> {console.log(<data-variable>)
// If error is thrown during runtime or while fetching;
use a catch block to avoid abnormalities in your webpage.
  .catch(err = > {console.err("Failed to fetch, error = ",
err) })
```

10.1.3.1 Code Snippet-1

```
// Store API URL as a constant.
const DOG_API ="https://dog.ceo/api/breeds/image/rand";
// Add event to button.
document.querySelector(".btn").addEventListener
("click", () => {
  fetch(DOG_API)
    // Wait for fetch to complete.
    .then((res) => {
      return res.json();
    })
    // Wait for data to be parsed into JSON format, which JS
understands.
    .then((apiData) => {
```

```
        // Convert JSON into string for displaying in web-
page (HTML can't display native JSON)
        document.querySelector(".info p").innerHTML =
JSON.stringify(apiData);
        // This line creates a reference to the img element,
and changes it's src value.
        document.querySelector(".doggo-here").src =
apiData.message;
        // Spilt the url with '/' as delimiter, and grab the
5th value, as it's the dog breed.
        // Look at the API's return message, to see the dog
breed, it's present in the link.
        document.querySelector(".breed-box").innerHTML
= apiData.message.split(
        "/"
        )[4];
    })
    .catch((err) => {
        console.log(err);
    });
});
```

10.2 Case Study-2: Integrating MongoDB in Cloud

10.2.1 Objective

i. Connecting and running some basic queries against MongoDB Atlas Cluster hosted in cloud (it is free) from a Node.js backend.

10.2.2 Prerequisite

10.2.2.1 MongoDB Cloud Connection

i. Create a MongoDB account here [4].

ii. Create a MongoDB Atlas cluster by following the on-screen instructions.

iii. Once created, click on collections and choose load sample dataset. This loads the sample data that you can play with.

iv. Now go to the connect tab and follow the on-screen instructions.

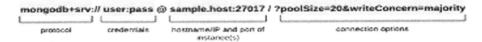

FIGURE 10.1
General breakdown of a connection string.

- When asked for IP, add "0.0.0.0/0" to avoid any runtime trouble. Use of this IP address is only recommended for the development stage.
- Copy the connection string, after clicking on **connect from application>** Breakdown of a connections string is given in Figure 10.1 [5].

10.2.2.2 Docker Container for Node.js

i. Install docker on your system (if you aren't familiar with Node.JS already).
ii. From https://www.routledge.com/9780367641429 download *Ch_10_mongodb*; under case_mongodb, you will get the codes required to execute this case study.
iii. You will find two new types of files: *Dockerfile* and *.dockerignore*. This case study requires a certain environment and runtime dependencies for which we will use these files.
iv. You don't need to worry about the runtime environment, as *Dockerfile* is there to define the environment.
v. Make changes to *server.js* and *public/script.js* to interact with JavaScript.
vi. Once you are ready to run, navigate to the project directory in *cmd* or *terminal*.
vii. Make sure that the file structure looks like the one shown in Figure 10.2.
   - Ignore *node_modules*; it's not required as of now.
   - Run these commands.
   - These commands must be run every time you make a change to any file in this case study.

```
<Feel free to replace "mongo-node" with any name you want>
docker build -t mongo-node .
docker run --name server -it -p 3000:3000 --rm mongo-node
```

FIGURE 10.2
Folder structure.

10.2.3 Agenda

- Creating a web page, which sends a request to our backend node.js server.
- A server can be reached by using two routes:
  - /databases -> returns JSON. (All the databases present in our MongoDB cluster).
  - /find -> takes URI encoded params and returns JSON. (All *hotels* with a given number of bedrooms, their address and description).

10.2.4 Explore

Using any editor, we will write and test JavaScript code to connect to and query against MongoDB cloud.

10.2.4.1 Front-End Script for Invoking Our Custom APIs

Put the following code inside the *public/script.js* file of your project directory.

```
  //DOM reference's
const btn = document.getElementById("btn");
const code = document.getElementById("code");
const search = document.getElementById("search");
const btnC = document.getElementById("btnC");
const db = document.getElementById("database");
const coll = document.getElementById("collection");

btn.addEventListener("click", async () => {
    code.innerText = "loading";

    const res = await fetch("/databases");
    const json = await res.json();
```

```
    code.innerText = "\n" + JSON.stringify(json, null, 4);
  });

  btnC.addEventListener("click", async () => {
      if (db.value) {
          const res = await fetch("/find?search = " +
encodeURIComponent(db.value));
          const json = await res.json();
          code.innerText  =  "\n"  +  JSON.stringify(json,
null, 4);
      } else {
          code.innerText =
           "Please provide a Database Name, hit 'Show data-
bases' to see available ones<3";
      }
  });
```

10.2.4.2 Code Walkthrough

- First section of the code demonstrates the access to DOM references.
- *fetch()* is a JavaScript API to fetch resources from different sources. It takes some input arguments: URL of the query, followed by params, which is optional.
- There are two fetch operations as needed for achieving aforementioned agenda. You can see *await* before fetch as fetch is acquiring information from different sources, and it is dependent on bandwidth as well. It will take some time before the data is available. *await* is used to halt the code execution till the process is complete and then to begin the code execution.

10.2.4.3 Creating Node.js Backend

Place the following code inside *server.js* file of the project directory.

```
// Official Mongo driver which is required to connect with
MongoDB clusters.

const { MongoClient } = require("mongodb");

// Framework for managing HTTP so we don't need to take
care of certain low-level things.
const express = require("express");
const { Router, response } = require("express");
```

```
const init = async () => {
    // TODO: Add your specific connection URI here.
    const uri =
        "ENTER URL FROM MONGODB CLOUD HERE";

    // Create a new instance of mongoclient by passing the
URI into constructor.
    const client = new MongoClient(uri, {
useUnifiedTopology: true });
    await client.connect();

    // Create a new instance of express, convenience dic-
tates to name it app.
    const app = express();
    // This is a server Route,
    // first parameter is the URL end-point which will
trigger this code snippet inside
    // A callback () is taken, with two parameters req->
request, and res-> response app.get("/databases", async
(req, res) => {
        // Fetching a list of all databases via the client
        const result = await
client.db().admin().listDatabases();

        // res (response) allows you to communicate back to the
client; first set the status code = 200 (This is optional, but a
good practice)
        // Followed by a JSON response, containing the result
acquired
        res.status(200).json({ dbs: result.databases }).end();
    });
    app.get("/find", async (req, res) => {
        const minBedrooms = Math.floor(req.query.search);
        console.log(minBedrooms);

        // This mongoDB query returns a cursor, which can be
further used to filter, sort, or access elements sequentially
        // We'll simply convert it into array to send all data
to client at once.
        // Here, $gte stands for GreaterThanEqualto, is one of
the many comparison operators in MongoDB, do check out the
official MongoDB documentation for more information.
        const cursor = await client
            .db("sample_airbnb")
            .collection("listingsAndReviews")
```

```
    .find({ bedrooms: { $gte: minBedrooms } })
    .sort({ last_review: -1 })
    .limit(50);

  const response = await cursor.toArray();
  var subset = [];

  // Extract selective sections of the result, else it's
a huge mess. I'm extracting name, description, and address.
    // Feel free to cosole.log(response) to see the entire
result, and manipulate accordingly.
    response.forEach((match) => {
        subset.push({
        name: match.name,
        dsc: match.description,
        addr: match.address,
      });
  });

  console.log(response);
  if (response) {
    res.
    status(200)
    .json({
        response:  `Valid  AirBnbs  with  ${minBedrooms}
minimum bedrooms`,
        matches: subset,
      })
      .end();
    } else {
     res.status(200).json({ match: "No matches found"
}).end();
    }
  });

  // Send the landing page
  // app.get("/", (req, res) => {
  // res.sendFile("public/index.html");
  // });

  // Initiate your server at PORT 3000. (Don't change the
port address, unless you've made the change in Docker, run
command. Ignore this if you have node installed on your PC and
running the code through it).
    const PORT = process.env.PORT || 3000;
```

```
     app.use(express.static("./public"));
     app.listen(PORT);

     console.log(`running on http://localhost:${PORT}`);
  };

  // Drive
  init();

  // Handle SIGINT
  process.on("SIGINT", () => {
     console.log("\nOkay, bye.");
     process.exit();
  });
```

</script>

10.3 Case Study-3: Visualizing Graphs with JavaScript

10.3.1 Objective

i. Visualizing graphs using random numbers that are generated in real time.

10.3.2 Prerequisite

i. From this https://www.routledge.com/9780367641429 download *Ch_10_visualizegraph*; you will get the codes required for this project.

10.3.3 Agenda

• Creating a web page that generates random floating-point numbers in fixed intervals and plotting them into the graph in real time.

10.3.4 Explore

Using any editor, we will write and test JavaScript code to plot and extend the graph in real time.

10.3.5 Front-End Script for Invoking Our Custom APIs

Place the following code inside *server.js* file of your project directory.

```
function randomInt(upperLimt = 500) {
    return Math.random() * upperLimt;
}

// plot method of plotly takes an argument, which is the ID
of DOM element.
    Plotly.plot("hook", [
        {
            // y: take a data array, of points to be plotted.
            y: [randomInt()],
            // Since we're plotting a line
            type: "line",
        },
    ]);

// Number of points that have been plotted.
// We will monitor this count, in order to extend the graph
when needed.
// By default X range is upto 500.
var x_points = 0;

// Set interval lets us execute a command with fixed in-
tervals in between. (30 ms in our example)
setInterval(function () {
    // The previously plotted graph will be extended, as the
same div-id is passed
    // randomInt() is called in every iteration and [0]
signifies 0th axis to be extended, as we have only one axis to
extend.
    Plotly.extendTraces("hook", { y: [[randomInt()]] }, [0]);
    x_points + +;

// Default range of x-axis is 500
// If the number of points plotted>500, it might extend
beyond the scope of graph, this function snippet keeps up-
dating the view to fit the graph as it extends.
    if (x_points>500) {
        Plotly.relayout("hook", {
            xaxis: {
    range: [x_points - 500, x_points],
            },
        });
    }
}, 30);
```

10.4 Case Study-4: Integrating Firebase Auth

10.4.1 Objective

i. Integrating firebase authentication into web app to allow login functionality by using Google's OAuth token (same can be extended to login using Apple ID, GitHub, Email/Pass etc.).

10.4.2 Background

Need for Authentication? (Simply taking an ID/Password in a textfield and a couple of if/else blocks won't work?)

- In a nutshell, no. There's much more to it. Basically, all login functionalities are built using two crucial subcomponents.
  - AUTHENTICATION that answers: Who is the user?
  - AUTHORIZATION that answers: What the user can do?

- AUTHORIZATION includes managing the state of user as in, Is it the same person, who got authenticated or is s/he still logged in? This is a very cumbersome server-side task, but firebase makes it really easy to manage and authorize users with its own back end. Luckily, we don't need to worry about that.

10.4.3 Prerequisite

i. Create a firebase project here, <u>Firebase Console.</u>
ii. Click on settings, next to project overview, and select **Project Settings.**
iii. Scroll down and click on **Add App.**
iv. Select **WebApp** and follow the on-screen instructions. Finally, you'll get the *firebaseConfig* object; keep it handy. You will need it for wiring the front end of our project with firebase.
v. From the left panel, click on **authentication** and select **get started**.
vi. Go to the **signin-method** tab on the **authentication** page and enable **google**. (You can avail any provider of choice; for this tutorial we're sticking to google.)
vii. Inside Firebase, **console -> Auth section -> Sign in** method tab, scroll down to **authorised-domains** and add your domain (most probably localhost or 127.0.0.1), in order to grant access to firebase.

viii. IMPORTANT* Your code will not work if you've opened the html file as a document; it needs to be served, like from localhost. One of the easier ways to do this is by using VisualStudio code editor. Just install **liveserver** extension, and use it to open html files.

10.4.4 Agenda

- Creating a login page through which users can log in using their google accounts.
- Creating a user page wherein post-login, the user's public credentials are displayed.

10.4.5 Explore

Using any editor, we will write and test JavaScript code to use Firebase authentication services and provide login functionality via the user's google account.

Resource:
- From https://www.routledge.com/9780367641429 download *Ch_10_FirebaseAuth*; you will get the code required for this project.
- Open up the *login.html* page, and you can see a couple of scripts with *src* from *gstatic*; it's a CDN (Content Delivery Network). As we have firebase as one of our dependency in this project, we will be using CDN to dynamically load firebase libraries into our runtime.

10.4.5.1 *Creating a Login Page*

Place the following code inside *script.js* file of your project directory.

```
var firebaseConfig = {
    /* Fill the firebaseConfig object with your project
credentials, as described in pre-requisites. */
    };

// Initialize Firebase
firebase.initializeApp(firebaseConfig);

// var provider = new firebase.auth.GoogleAuthProvider();

var uiConfig = {
    signInSuccessUrl: "home.html",
    signInOptions: [
```

```
        // Leave the lines as it is for the providers you
want to offer your users.
            firebase.auth.GoogleAuthProvider.PROVIDER_ID,
        ],
        // tosUrl and privacyPolicyUrl accept either url
string or a callback function.
        // Terms of service url/callback.
        tosUrl: "https://www.google.co.in",
        // Privacy policy url/callback.
        privacyPolicyUrl: function () {
            window.location.assign("https://www.google.co.in");
        },
    };

    // Initialize the FirebaseUI Widget using Firebase.
    var ui = new firebaseui.auth.AuthUI(firebase.auth());
    // The start method will wait until the DOM is loaded.
    ui.start("#firebaseui-auth-container", uiConfig);
```

10.4.5.2 Code Walkthrough

- *firebaseConfig* basically allows you to connect to the firebase service with credentials.
- *uiConfig* takes input as two of the important attributes; these are mentioned as follows:
  - *signInSuccessUrl*: URL of the page where the control is to be transferred upon successful authentication.
  - *signInOptions*: An array of the different ways for signing in are mentioned. As we are using google here, we just need one element. (Whatever methods need to be implemented, make sure, it's enabled in the firebase console.)
- New *firebaseui.auth.AuthUI(firebase.auth())* is the constructor that initiates the authentication flow. We are using *AuthUI*, as it manages the login with Google's standard UI elements.
- Finally *ui.start()* takes the *selector*, either ID or CLASS of the *div* where you need to hook the login portal. Our HTML page already consists of div with *id = "firebaseui-auth-container"*. Second, start() also takes the *uiconfig*, and it starts working only when the entire web page is loaded.

10.4.5.3 *Creating Home Page*

This page will be automatically redirected as soon as login is successful, as described in the previous snippet. Place the following code snippet inside *auth.js* file of your project directory.

```javascript
var firebaseConfig = {
    /* TODO:
     * Populate four firebase project credentials here. <3
     */
};

// Initialize Firebase
firebase.initializeApp(firebaseConfig);

// DOM references
var fname = document.querySelector(".photo-details h1");
var dp = document.querySelector(".photo-card img");
const card = document.querySelector(".photo-card");

var userEmail, userPicURL, userName;

const logout = document.getElementById("logout");
initApp = function () {
  // Real time listener.
  firebase.auth().onAuthStateChanged(
      (user) => {
            if (user) {
                // User authenticated and logged in.
                card.classList.remove("hide");
                populateLoggedInUserCard(user);
//Login card
                logout.classList.remove("hide");
//Unhide logout button
                userEmail = user.email;
                userPicURL = user.photoURL;
                userName = user.displayName;
            } else {
                // Hide logout button
                // logout.classList.add("hide");
                card.classList.add("hide");

                userEmail = null;
                userPicURL = null;
                userName = null;
```

```
        }
    },
    (error) => {
        console.log(error);
    });

    // log-out event listener.
    document.getElementById("logout").addEventListener
("click", (e) => {
            firebase.
            auth().
            signOut().
            then(() => {
                // Sign-out successful.
                window.location.replace("login.html");

            })
            .catch((error) => {
                // An error happened.
            });
    });
};

// End of log-out event listener.

// Hold back initialization until DOM is loaded.
window.addEventListener("load", function () {
    initApp();
});

// Populate logged-in user card.
const populateLoggedInUserCard = (user) => {
    fname.innerText = user.displayName;
    dp.src = user.photoURL;
};
```

10.4.5.4 Code Walkthrough

This code is pretty much self-descriptive in its nature with all the comments, and here is a quick walkthrough of some the key elements that are used to make it work.

- firebase.auth().onAuthStateChanged(...)
  - Normally we would use a promise attached to a login func-
    tionality, but that would fire only once after login is successful

and a lot of complex codes to handle that dynamically. Firebase provides onAuthStateChanged() method that is triggered every time the user logs in or logs out and takes a callback to be executed in that event (with a parameter *user* in our code snippet). It will be *undefined/null* when logout event takes place; else it will be holding public credentials of the user, as passed on by the OAuth token. Try printing *user* variable into the console, to see for yourself what data is provided.

    o Unique identifier (UID) provided within the OAuth data is guaranteed to be unique across all the users within the same firebase project. Hence, it is safe to use for selective database access or rules. Although that's not a part of this case study but is the most crucial component if you plan to do something related to authorization.

To log out, you just need to invoke the *firebase.auth().signOut()* method, which resolves into a promise; hence, it is followed by a *then()* to chain the promise and a *catch()* block in case the promise is not handled properly.

References

[1] "HTML: HyperText Markup Language | MDN." [Online]. Available: https://developer.mozilla.org/en-US/docs/Web/HTML. [Accessed: 31-May-2021].

[2] "Cascading Style Sheets." [Online]. Available: https://www.w3.org/Style/CSS/Overview.en.html. [Accessed: 31-May-2021].

[3] "JavaScript | MDN." [Online]. Available: https://developer.mozilla.org/en-US/docs/Web/JavaScript. [Accessed: 31-May-2021].

[4] "Difference between Client side JavaScript and Server side JavaScript." [Online]. Available: http://net-informations.com/js/iq/side.htm. [Accessed: 31-May-2021].

[5] "About | Node.js." [Online]. Available: https://nodejs.org/en/about/. [Accessed: 31-May-2021].

[6] "React – A JavaScript Library for Building User Interfaces." [Online]. Available: https://reactjs.org/. [Accessed: 31-May-2021].

[7] "AngularJS — Superheroic JavaScript MVW Framework." [Online]. Available: https://angularjs.org/. [Accessed: 31-May-2021].

[8] "The History of JavaScript: A Journey from Netscape to Frameworks." [Online]. Available: https://www.techaheadcorp.com/blog/history-of-javascript/. [Accessed: 31-May-2021].

[9] "LiveScript - A Language which Compiles to JavaScript." [Online]. Available: https://livescript.net/. [Accessed: 31-May-2021].

[10] "Getting Started with JavaScript – The Right Tools and Resources (Video Interview) | Christian Heilmann." [Online]. Available: https://christianheilmann.com/2019/04/18/getting-started-with-javascript-the-right-tools-and-resources-video-interview/. [Accessed: 31-May-2021].

[11] G. Repository et al. "ECMAScript ® 2019 language specification," June, pp. 1–764, 2019.

[12] "ECMA-262 - Ecma International." [Online]. Available: https://www.ecma-international.org/publications-and-standards/standards/ecma-262/. [Accessed: 31-May-2021].

[13] "jQuery." [Online]. Available: https://jquery.com/. [Accessed: 31-May-2021].

[14] "Dojo Toolkit." [Online]. Available: https://dojotoolkit.org/. [Accessed: 31-May-2021].

[15] "TC39 - Specifying JavaScript." [Online]. Available: https://tc39.es/. [Accessed: 31-May-2021].

[16] "DOM Living Standard," *Dom specification*. [Online]. Available: https://dom.spec.whatwg.org/. [Accessed: 20-Apr-2021].

[17] "Firefox - Protect your life online with privacy-first products — Mozilla." [Online]. Available: https://www.mozilla.org/en-US/firefox/. [Accessed: 31-May-2021].

[18] "Internet Explorer Downloads." [Online]. Available: https://support.microsoft.com/en-us/windows/internet-explorer-downloads-d49e1f0d-571c-9a7b-d97e-be248806ca70. [Accessed: 31-May-2021].

[19] "Safari - Apple (IN)." [Online]. Available: https://www.apple.com/in/safari/. [Accessed: 31-May-2021].

[20] "Google Chrome Web Browser." [Online]. Available: https://www.google.com/intl/en_in/chrome/. [Accessed: 31-May-2021].

[21] "Netscape Navigator - MDN Web Docs Glossary: Definitions of Web-related terms | MDN." [Online]. Available: https://developer.mozilla.org/en-US/docs/Glossary/Netscape_Navigator. [Accessed: 31-May-2021].

[22] "MSHTML Reference (Internet Explorer) | Microsoft Docs.".

[23] "Gecko:Home Page - MozillaWiki." [Online]. Available: https://wiki.mozilla.org/Gecko:Home_Page. [Accessed: 31-May-2021].

[24] "Blink (Rendering Engine) - The Chromium Projects." [Online]. Available: https://www.chromium.org/blink. [Accessed: 31-May-2021].

[25] "WebKit." [Online]. Available: https://webkit.org/. [Accessed: 31-May-2021].

[26] "Window.alert() - Web APIs | MDN." [Online]. Available: https://developer.mozilla.org/en-US/docs/Web/API/Window/alert. [Accessed: 31-May-2021].

[27] "Interaction: Alert, Prompt, Confirm." [Online]. Available: https://javascript.info/alert-prompt-confirm. [Accessed: 31-May-2021].

[28] "Window.prompt() - Web APIs | MDN." [Online]. Available: https://developer.mozilla.org/en-US/docs/Web/API/Window/prompt. [Accessed: 31-May-2021].

[29] "Document.write() - Web APIs | MDN." [Online]. Available: https://developer.mozilla.org/en-US/docs/Web/API/Document/write. [Accessed: 31-May-2021].

[30] "JavaScript | console.log() with Examples - GeeksforGeeks." [Online]. Available: https://www.geeksforgeeks.org/javascript-console-log-with-examples/. [Accessed: 31-May-2021].

[31] "Express - Node.js web application framework." [Online]. Available: https://expressjs.com/. [Accessed: 31-May-2021].

[32] "Electron | Build cross-platform desktop apps with JavaScript, HTML, and CSS." [Online]. Available: https://www.electronjs.org/. [Accessed: 31-May-2021].

[33] "NPM for JavaScript - ' Nocturnally Psychologizing Millipede.'" [Online]. Available: https://www.npmjs.com/. [Accessed: 31-May-2021].

[34] "Gulp Toolkit for JavaScript." [Online]. Available: https://gulpjs.com/. [Accessed: 31-May-2021].

[35] "Webpack5 for JavaScript." [Online]. Available: https://webpack.js.org/. [Accessed: 31-May-2021].

[36] "Vue.js." [Online]. Available: https://vuejs.org/. [Accessed: 31-May-2021].

[37] "jQuery Tutorial: Using a JavaScript Library | Tania Rascia." [Online]. Available: https://www.taniarascia.com/how-to-use-jquery-a-javascript-library/. [Accessed: 31-May-2021].

[38] "Lexical Structure - JavaScript: The Definitive Guide, 7th Edition [Book]." [Online]. Available: https://www.oreilly.com/library/view/javascript-the-definitive/9781491952016/ch02.html. [Accessed: 31-May-2021].

[39] "Character Set - JavaScript: The Definitive Guide, 6th Edition [Book]." [Online]. Available: https://www.oreilly.com/library/view/javascript-the-definitive/9781449393854/ch02s01.html. [Accessed: 31-May-2021].

[40] "HTML ASCII Reference." [Online]. Available: https://www.w3schools.com/charsets/ref_html_ascii.asp. [Accessed: 31-May-2021].

[41] "Identifier - MDN Web Docs Glossary: Definitions of Web-related terms | MDN." [Online]. Available: https://developer.mozilla.org/en-US/docs/Glossary/Identifier. [Accessed: 31-May-2021].

[42] "JavaScript Reserved Words." [Online]. Available: https://www.w3schools.com/js/js_reserved.asp. [Accessed: 31-May-2021].

[43] "JavaScript data types and data structures - JavaScript | MDN." [Online]. Available: https://developer.mozilla.org/en-US/docs/Web/JavaScript/Data_structures. [Accessed: 31-May-2021].

[44] K. Simpson, *You Don't Know JS: Scope & Closures*. O'Reilly Media, Inc., 2014.

[45] "Primitive - MDN Web Docs Glossary: Definitions of Web-related terms | MDN." [Online]. Available: https://developer.mozilla.org/en-US/docs/Glossary/Primitive. [Accessed: 31-May-2021].

[46] "String - JavaScript | MDN." [Online]. Available: https://developer.mozilla.org/en-US/docs/Web/JavaScript/Reference/Global_Objects/String. [Accessed: 31-May-2021].

[47] D. Crockford, *How JavaScript Works*. Virgule-Solidus, 2018.

[48] "Boolean - JavaScript | MDN." [Online]. Available: https://developer.mozilla.org/en-US/docs/Web/JavaScript/Reference/Global_Objects/Boolean. [Accessed: 31-May-2021].

[49] MDN Web Docs, "BigInt." [Online]. Available: https://developer.mozilla.org/en-US/docs/Glossary/BigInt. [Accessed: 18-May-2021].

[50] "Symbol - JavaScript | MDN." [Online]. Available: https://developer.mozilla.org/en-US/docs/Web/JavaScript/Reference/Global_Objects/Symbol. [Accessed: 31-May-2021].

[51] "undefined - JavaScript | MDN." [Online]. Available: https://developer.mozilla.org/en-US/docs/Web/JavaScript/Reference/Global_Objects/undefined. [Accessed: 31-May-2021].

[52] "null - JavaScript | MDN." [Online]. Available: https://developer.mozilla.org/en-US/docs/Web/JavaScript/Reference/Global_Objects/null. [Accessed: 31-May-2021].

[53] "Typecasting and Coercion in JavaScript | by Aquil Hussain | The Startup | Medium." [Online]. Available: https://medium.com/swlh/typecasting-and-coercion-in-javascript-f0d59b0a86db. [Accessed: 31-May-2021].

[54] "Expressions and Operators - JavaScript | MDN." [Online]. Available: https://developer.mozilla.org/en-US/docs/Web/JavaScript/Guide/Expressions_and_Operators. [Accessed: 31-May-2021].

[55] "JavaScript | Arithmetic Operators - GeeksforGeeks." [Online]. Available: https://www.geeksforgeeks.org/javascript-arithmetic-operators/. [Accessed: 31-May-2021].

[56] "Unicode in JavaScript." [Online]. Available: https://flaviocopes.com/javascript-unicode/. [Accessed: 31-May-2021].

[57] "Bitwise AND (&) - JavaScript | MDN." [Online]. Available: https://developer.mozilla.org/en-US/docs/Web/JavaScript/Reference/Operators/Bitwise_AND. [Accessed: 31-May-2021].

[58] "JavaScript Comparison and Logical Operators." [Online]. Available: https://www.w3schools.com/js/js_comparisons.asp. [Accessed: 31-May-2021].

[59] "Conditional (ternary) operator - JavaScript | MDN." [Online]. Available: https://developer.mozilla.org/en-US/docs/Web/JavaScript/Reference/Operators/Conditional_Operator. [Accessed: 31-May-2021].

[60] "Comma operator (,) - JavaScript | MDN." [Online]. Available: https://developer.mozilla.org/en-US/docs/Web/JavaScript/Reference/Operators/Comma_Operator. [Accessed: 31-May-2021].

[61] "typeof - JavaScript | MDN." [Online]. Available: https://developer.mozilla.org/en-US/docs/Web/JavaScript/Reference/Operators/typeof. [Accessed: 31-May-2021].

[62] MDN Web Docs, "No Title." [Online]. Available: https://developer.mozilla.org/en-US/docs/Web/JavaScript/Reference/Operators/typeof. [Accessed: 21-May-2021].

[63] "delete operator - JavaScript | MDN." [Online]. Available: https://developer.mozilla.org/en-US/docs/Web/JavaScript/Reference/Operators/delete. [Accessed: 31-May-2021].

[64] "void operator - JavaScript | MDN." [Online]. Available: https://developer.mozilla.org/en-US/docs/Web/JavaScript/Reference/Operators/void. [Accessed: 31-May-2021].

[65] "Control Flow - MDN Web Docs Glossary: Definitions of Web-related terms | MDN." [Online]. Available: https://developer.mozilla.org/en-US/docs/Glossary/Control_flow. [Accessed: 31-May-2021].

[66] "Loops and Iteration - JavaScript | MDN." [Online]. Available: https://developer.mozilla.org/en-US/docs/Web/JavaScript/Guide/Loops_and_iteration. [Accessed: 31-May-2021].

[67] "Working with Objects - JavaScript | MDN." [Online]. Available: https://developer.mozilla.org/en-US/docs/Web/JavaScript/Guide/Working_with_Objects. [Accessed: 31-May-2021].

[68] D. A. Rauschmayer, *JavaScript for Impatient Programmers*. Independently Published (30 August 2019) ISBN-10 : 1091210098USA.

[69] D. Cosset, "Maps in Javascript ES6," *dev.to*, 2018. [Online]. Available: https://dev.to/damcosset/maps-in-javascript-es6-4301. [Accessed: 14-May-2021].

[70] D. A. Rauschmayer, "Objects as Dictionaries (Advanced)," 2019. [Online]. Available: https://exploringjs.com/impatient-js/ch_single-objects.html#objects-as-dictionaries-advanced. [Accessed: 20-Apr-2021].

[71] M. Gupta, "The Ultimate Guide to the JavaScript Delete Keyword," *Medium.com*, 2015. [Online]. Available: https://medium.com/technofunnel/javascript-delete-keyword-in-detail-4bdcf32dcdd8.

[72] "Inheritance and the Prototype Chain - JavaScript | MDN." [Online]. Available: https://developer.mozilla.org/en-US/docs/Web/JavaScript/Inheritance_and_the_prototype_chain. [Accessed: 31-May-2021].

[73] K. Simpson, *You Don't Know JS: ES6 & Beyond*. O'Reilly Media, Inc., 2015.

[74] "Object - JavaScript | MDN." [Online]. Available: https://developer.mozilla.org/en-US/docs/Web/JavaScript/Reference/Global_Objects/Object. [Accessed: 31-May-2021].

[75] "static - JavaScript | MDN." [Online]. Available: https://developer.mozilla.org/en-US/docs/Web/JavaScript/Reference/Classes/static. [Accessed: 31-May-2021].

[76] "JavaScript Function Definitions." [Online]. Available: https://www.w3schools.com/js/js_function_definition.asp. [Accessed: 31-May-2021].

[77] "Functions - JavaScript | MDN." [Online]. Available: https://developer.mozilla.org/en-US/docs/Web/JavaScript/Guide/Functions. [Accessed: 31-May-2021].

[78] "Function Expression - JavaScript | MDN." [Online]. Available: https://developer.mozilla.org/en-US/docs/Web/JavaScript/Reference/Operators/function. [Accessed: 31-May-2021].

[79] "Arrow Function Expressions - JavaScript | MDN." [Online]. Available: https://developer.mozilla.org/en-US/docs/Web/JavaScript/Reference/Functions/Arrow_functions. [Accessed: 31-May-2021].

[80] "Recursive Functions in JavaScript | by Ross Mawdsley | The Startup | Medium." [Online]. Available: https://medium.com/swlh/recursive-functions-in-javascript-9ab0dd97e486. [Accessed: 31-May-2021].

[81] "Closures - JavaScript | MDN." [Online]. Available: https://developer.mozilla.org/en-US/docs/Web/JavaScript/Closures. [Accessed: 31-May-2021].

[82] "Array - JavaScript | MDN." [Online]. Available: https://developer.mozilla.org/en-US/docs/Web/JavaScript/Reference/Global_Objects/Array. [Accessed: 31-May-2021].

[83] "Learn JavaScript Multidimensional Array By Examples." [Online]. Available: https://www.javascripttutorial.net/javascript-multidimensional-array/. [Accessed: 31-May-2021].

[84] "JavaScript - BOM (Browser Object Model) | javascript Tutorial." [Online]. Available: https://riptutorial.com/javascript/topic/3986/bom--browser-object-model-. [Accessed: 31-May-2021].

[85] "Window - Web APIs | MDN." [Online]. Available: https://developer.mozilla.org/en-US/docs/Web/API/Window. [Accessed: 31-May-2021].

[86] "History - Web APIs | MDN." [Online]. Available: https://developer.mozilla.org/en-US/docs/Web/API/History. [Accessed: 31-May-2021].

[87] "Navigator - Web APIs | MDN." [Online]. Available: https://developer.mozilla.org/en-US/docs/Web/API/Navigator. [Accessed: 31-May-2021].

[88] "Location - Web APIs | MDN." [Online]. Available: https://developer.mozilla.org/en-US/docs/Web/API/Location. [Accessed: 31-May-2021].

[89] "Window.screen - Web APIs | MDN." [Online]. Available: https://developer.mozilla.org/en-US/docs/Web/API/Window/screen. [Accessed: 31-May-2021].

[90] "Document Object Model (DOM) - Web APIs | MDN." [Online]. Available: https://developer.mozilla.org/en-US/docs/Web/API/Document_Object_Model. [Accessed: 31-May-2021].

[91] "DOM & BOM Revisited. Heads up! This is a cross-post of... | by Federico Knüssel | Medium." [Online]. Available: https://medium.com/@fknussel/dom-bom-revisited-cf6124e2a816. [Accessed: 31-May-2021].

[92] "How to Create a DOM Tree - Web APIs | MDN." [Online]. Available: https://developer.mozilla.org/en-US/docs/Web/API/Document_object_model/How_to_create_a_DOM_tree. [Accessed: 31-May-2021].

[93] R. Bohdanowicz, "D3.DOM Visualizer," *BIOUB*. [Online]. Available: http://bioub.github.io/dom-visualizer/. [Accessed: 25-May-2021].

[94] "Node Web APIs | MDN." [Online]. Available: https://developer.mozilla.org/en-US/docs/Web/API/Node. [Accessed: 31-May-2021].

[95] "Document - Web APIs | MDN." [Online]. Available: https://developer.mozilla.org/en-US/docs/Web/API/Document. [Accessed: 31-May-2021].

[96] "Element - Web APIs | MDN." [Online]. Available: https://developer.mozilla.org/en-US/docs/Web/API/Element. [Accessed: 31-May-2021].

[97] "Introduction to Events - Learn Web Development | MDN." [Online]. Available: https://developer.mozilla.org/en-US/docs/Learn/JavaScript/Building_blocks/Events. [Accessed: 31-May-2021].

[98] "Error - JavaScript | MDN." [Online]. Available: https://developer.mozilla.org/en-US/docs/Web/JavaScript/Reference/Global_Objects/Error. [Accessed: 31-May-2021].

[99] "Number - JavaScript | MDN." [Online]. Available: https://developer.mozilla.org/en-US/docs/Web/JavaScript/Reference/Global_Objects/Number. [Accessed: 31-May-2021].

[100] "Unix Time," *Wikipedia*. [Online]. Available: https://en.wikipedia.org/wiki/Unix_time. [Accessed: 12-May-2021].

[101] MDN Web Docs, "Date.now()." [Online]. Available: https://developer.mozilla.org/en-US/docs/Web/JavaScript/Reference/Global_Objects/Date/now. [Accessed: 17-May-2021].

[102] MDN Web Docs, "Date.prototype.toISOString()," *Mozilla*. [Online]. Available: https://developer.mozilla.org/en-US/docs/Web/JavaScript/Reference/Global_Objects/Date/toISOString. [Accessed: 10-May-2021].

[103] "Math - JavaScript | MDN." [Online]. Available: https://developer.mozilla.org/en-US/docs/Web/JavaScript/Reference/Global_Objects/Math. [Accessed: 31-May-2021].

[104] "RegExp - JavaScript | MDN." [Online]. Available: https://developer.mozilla.org/en-US/docs/Web/JavaScript/Reference/Global_Objects/RegExp. [Accessed: 31-May-2021].

[105] "Regular Expressions - JavaScript | MDN." [Online]. Available: https://developer.mozilla.org/en-US/docs/Web/JavaScript/Guide/Regular_Expressions. [Accessed: 31-May-2021].

[106] "Keyed Collections - JavaScript | MDN." [Online]. Available: https://developer.mozilla.org/en-US/docs/Web/JavaScript/Guide/Keyed_collections. [Accessed: 31-May-2021].

[107] "Map - JavaScript | MDN." [Online]. Available: https://developer.mozilla.org/en-US/docs/Web/JavaScript/Reference/Global_Objects/Map. [Accessed: 31-May-2021].

[108] "Set - JavaScript | MDN." [Online]. Available: https://developer.mozilla.org/en-US/docs/Web/JavaScript/Reference/Global_Objects/Set. [Accessed: 31-May-2021].

[109] "Indexed collections - JavaScript | MDN." [Online]. Available: https://developer.mozilla.org/en-US/docs/Web/JavaScript/Guide/Indexed_collections. [Accessed: 31-May-2021].

[110] S. S. Sriparasa, *JavaScript and JSON Essentials*. Packt Publishing Ltd., 2013.

[111] MDN, "Client-side form Validation." Available: https://developer.mozilla.org/en-US/docs/Learn/Forms/Form_validation

[112] "Server and Client Side Validation with JavaScript, HTML, and Hapi." [Online]. Available: https://medium.com/@davidpetri/server-and-client-side-validation-with-javascript-html-and-hapi-js-eccc779e448a. [Accessed: 15-May-2021].

[113] "Constraint Validation API." [Online]. Available: https://developer.mozilla.org/en-US/docs/Web/API/Constraint_validation. [Accessed: 10-May-2021].

[114] "try...catch - JavaScript | MDN." [Online]. Available: https://developer.mozilla.org/en-US/docs/Web/JavaScript/Reference/Statements/try...catch. [Accessed: 31-May-2021].

[115] "throw - JavaScript | MDN." [Online]. Available: https://developer.mozilla.org/en-US/docs/Web/JavaScript/Reference/Statements/throw. [Accessed: 31-May-2021].

[116] "Document.cookie - Web APIs | MDN." [Online]. Available: https://developer.mozilla.org/en-US/docs/Web/API/Document/cookie. [Accessed: 31-May-2021].

[117] MDN, "Strict Mode." [Online]. Available: https://developer.mozilla.org/en-US/docs/Web/JavaScript/Reference/Strict_mode. [Accessed: 15-May-2021].

[118] "Using Fetch - Web APIs | MDN." [Online]. Available: https://developer.mozilla.org/en-US/docs/Web/API/Fetch_API/Using_Fetch. [Accessed: 31-May-2021].

[119] "Dog API." [Online]. Available: https://dog.ceo/dog-api/. [Accessed: 31-May-2021].

[120] "15 Fun APIs For Your Next Project - DEV Community." [Online]. Available: https://dev.to/biplov/15-fun-apis-for-your-next-project-5053. [Accessed: 31-May-2021].

[121] "Log in | MongoDB." [Online]. Available: https://account.mongodb.com/account/login?n=%2Fv2%2F6090eb552291122c0b36e2fb%23metrics%

2FreplicaSet%2F60b14aae4caa2f37265130da%2Fexplorer%2Fsample_airbnb%2FlistingsAndReviews%2Ffind. [Accessed: 31-May-2021].

[122] "Connection Guide – Node.js." [Online]. Available: https://docs.mongodb.com/drivers/node/current/fundamentals/connection/. [Accessed: 31-May-2021].

Index

Printed in the United States
by Baker & Taylor Publisher Services